D0516361

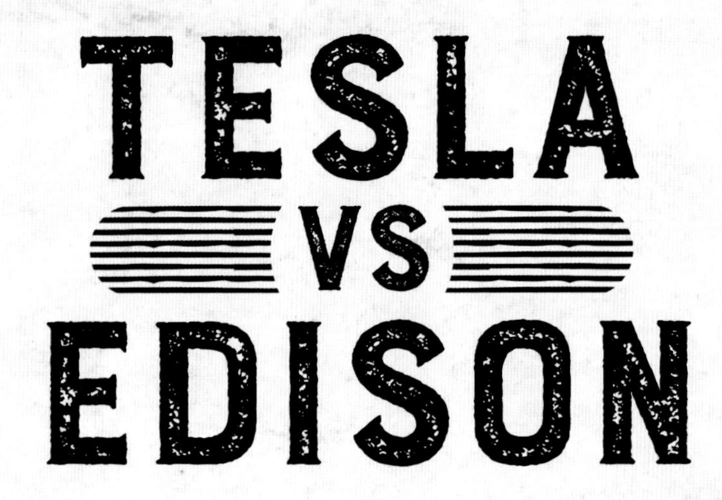
TESLA
VS
EDISON

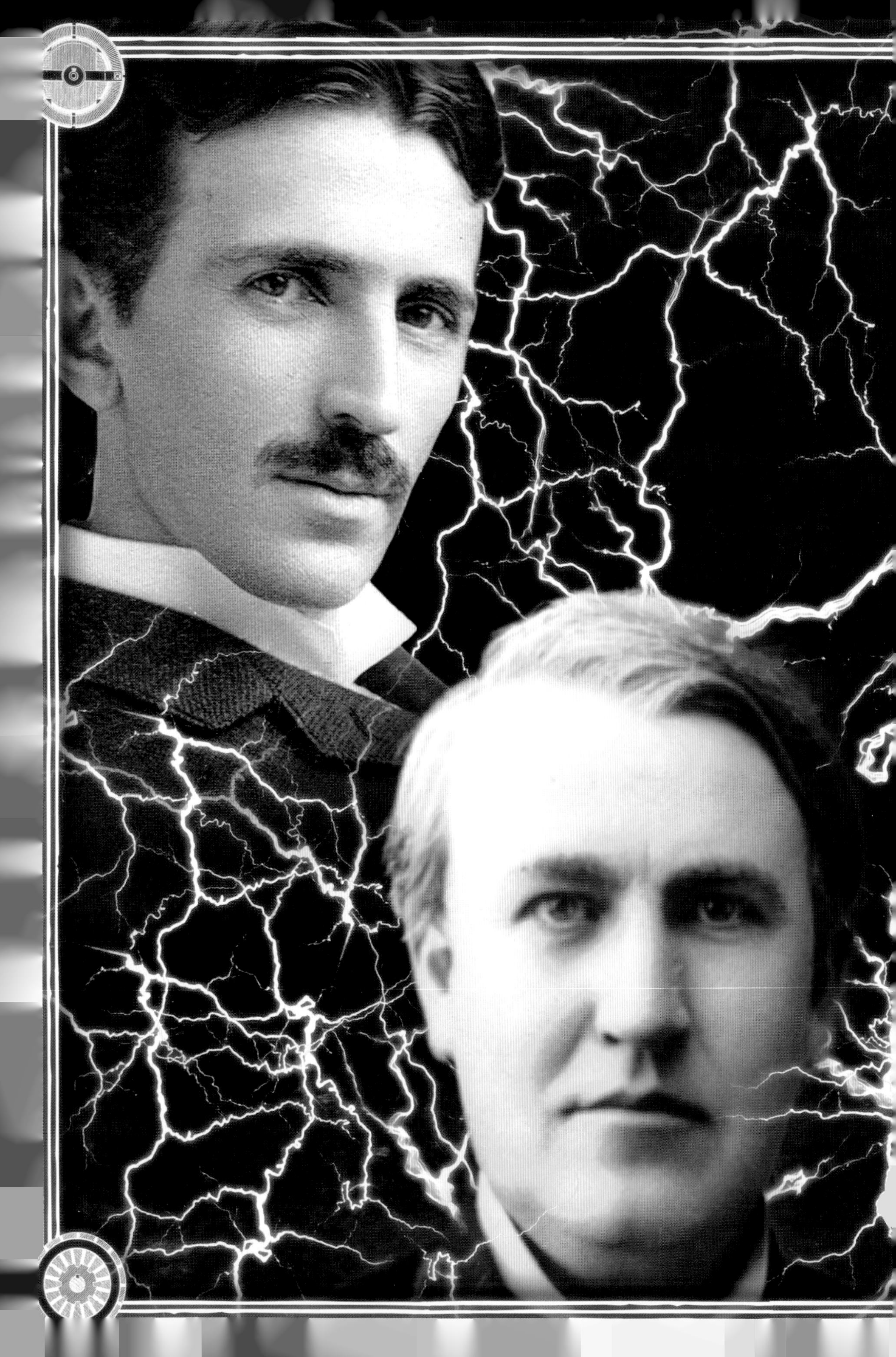

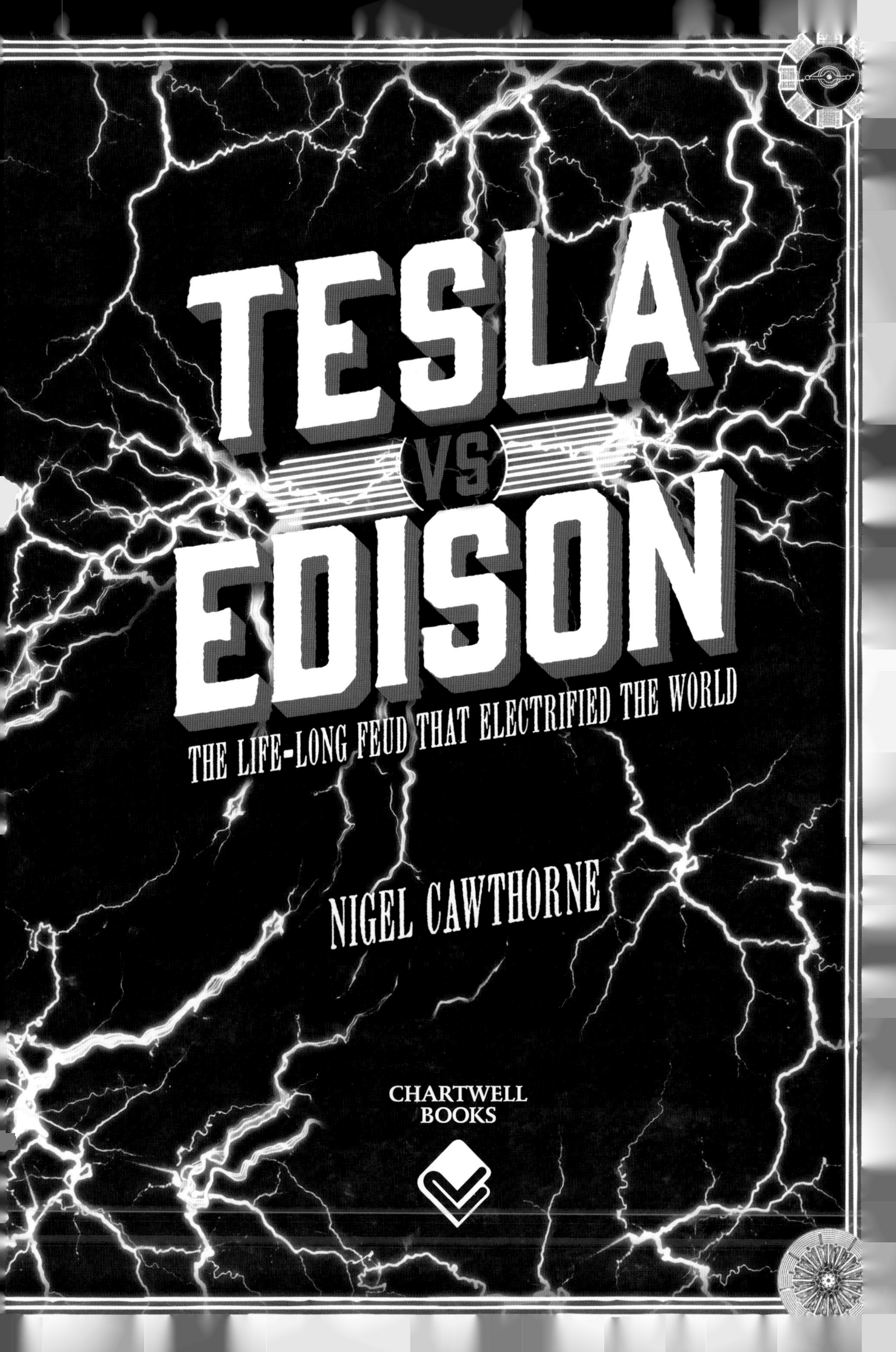
TESLA
VS
EDISON
THE LIFE-LONG FEUD THAT ELECTRIFIED THE WORLD
NIGEL CAWTHORNE
CHARTWELL
BOOKS

None of my inventions came by accident. What it boils down to is one percent inspiration and ninety-nine percent perspiration.

Thomas Edison
(1847 - 1931)

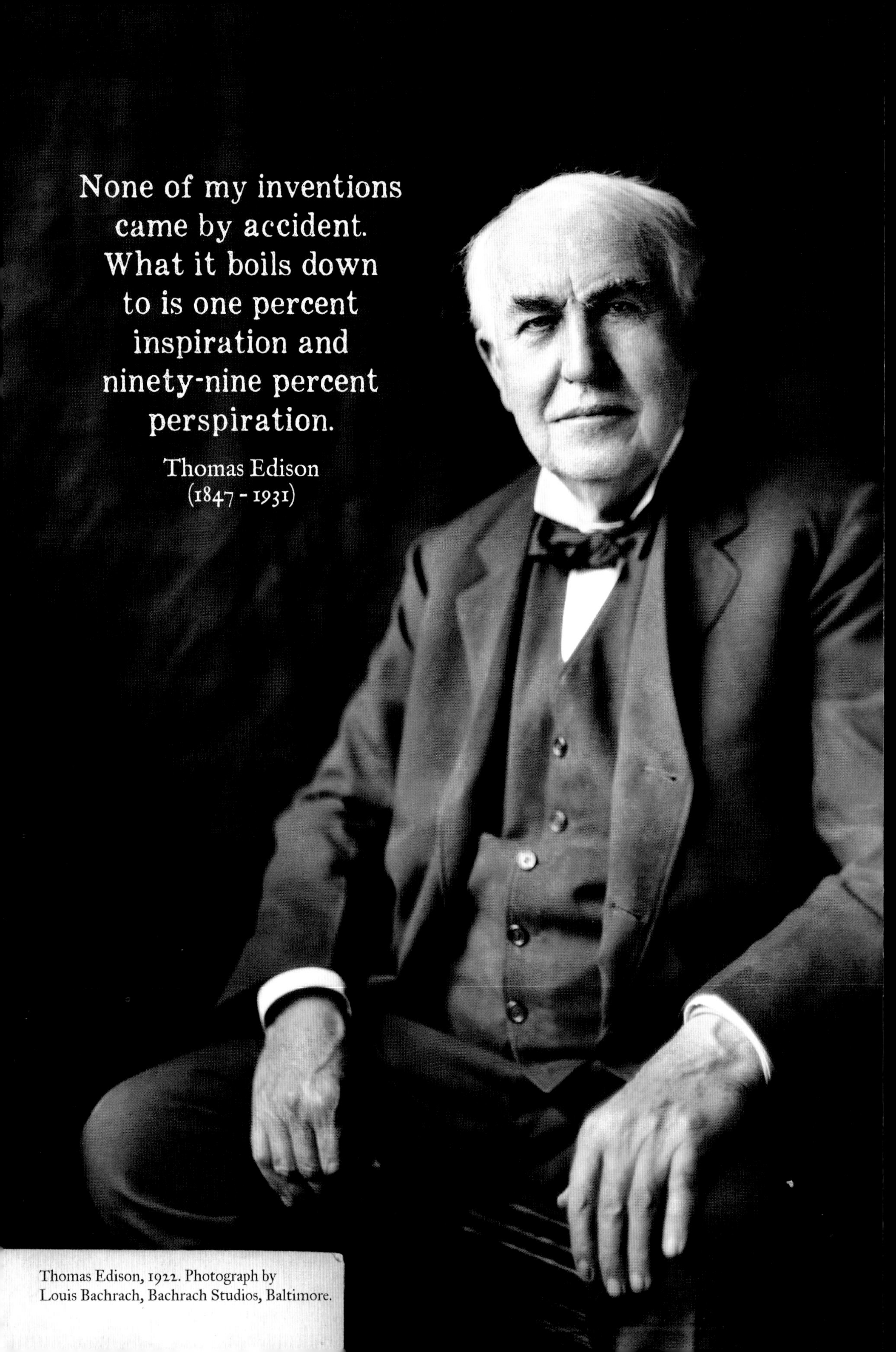

Thomas Edison, 1922. Photograph by
Louis Bachrach, Bachrach Studios, Baltimore.

CONTENTS

INTRODUCTION

I was thrilled to the marrow by meeting Edison who began my American education right then and there. I wanted to have my shoes shined, something I considered below my dignity. Edison said: "Tesla, you will shine the shoes yourself and like it." He impressed me tremendously. I shined my shoes and liked it.

Nikola Tesla[1]

That is how Tesla recalled his meeting with Thomas Edison in 1884, in his autobiography published thirty-five years later. However, in Edison's official biography, *Edison: His Life and Inventions,* published in 1910, Tesla gets just one mention. The passage concerns the establishing of the French Edison Company at Ivry sur-Seine, near Paris, in 1882 and says: "It was there that Mr. Nikola Tesla made his entree into the field of light and power, and began his own career as an inventor ..." [2]

There is no mention of Tesla's work as an employee of Edison in New York two years later. Nor is there any mention of Tesla as Edison's great rival in the "battle of the currents" where the two men's electrical systems came into direct competition in the struggle to electrify America – and ultimately the world. Tesla won out and it is Tesla's system we use to this day.

In their time, they were both famous. But Tesla, a truly original mind, is largely forgotten, while Edison, with his genius for improving other people's inventions and unleashing them on the world, still has his name up in lights. Literally. He is

Nikola Tesla reading in front of the spiral coil of his high-frequency transformer at his East Houston Street laboratory in New York.

remembered, of course, as the inventor of the light bulb – though it could be argued that he was the first among equals when it came to its invention.

Edison is remembered because he was a businessman who could exploit the rapid scientific advances that were being made in the field of electromagnetism at the turn of the twentieth century. The light bulb rapidly banished the darkness across the civilized world, thanks to Edison's entrepreneurial spirit. He was also a brilliant self-publicist. This is not to diminish his technical abilities, which even Tesla recognized.

By contrast, Tesla was a visionary. But without the backing of the great entrepreneur and gifted engineer George Westinghouse, Tesla's revolutionary inventions would probably have come to nothing.

When Edison died in 1931, at his mansion *Glenmont* in New Jersey, he had over a thousand patents to his name and was burdened with accolades, including the *Légion d'Honneur* and the Congressional Gold Medal. Since then, his birthday, February 11, has been designated National Inventor's Day. In 1997, *Life* magazine named him one of the "100 Most Important People in the Last 1,000 Years," and in 2010, he was given a Grammy, for his technical contribution to recorded sound. He invented the phonograph, the forerunner of the gramophone.[3]

Tesla died alone and penniless, in a hotel room in New York in 1943. A few months later, the US Supreme Court ruled that a number of patents claimed by Marconi actually belonged to Tesla. He was the unacknowledged father of radio – another of the many inventions we owe to him.[4]

Since then, there have been further attempts to get Tesla the recognition he is due. He may have won the "battle of the currents," but in the fame stakes Edison is the clear front-runner. But perhaps the battle between the two men still has a long way to run.

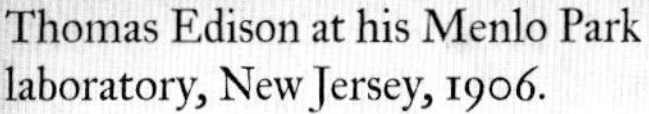

Thomas Edison at his Menlo Park laboratory, New Jersey, 1906.

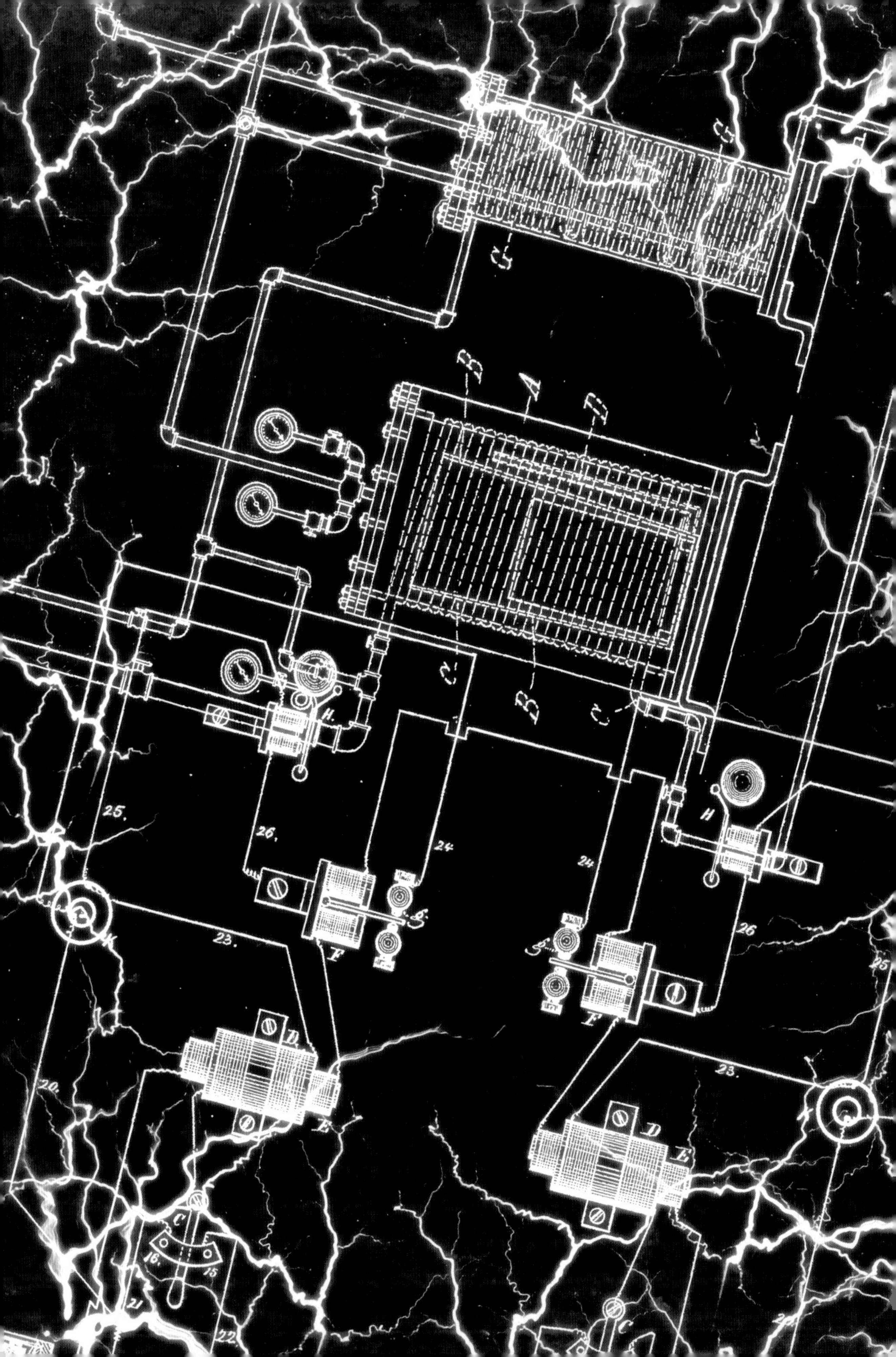

A
B
C
H.
H
25.
26.
24
24
26
23.
23.
F
F
D
E
D
E
25
20.
C
16.
15
21
22.

PART ONE

THE WIZARD OF MENLO PARK

CHAPTER 1

ENCOUNTERING THOMAS EDISON

I was amazed at this wonderful man who, without early advantages and scientific training, had accomplished so much. I had studied a dozen languages, delved in literature and art, and had spent my best years in libraries reading all sorts of stuff that fell into my hands, from Newton's Principia *to the novels of* [long forgotten trashy French writer] *Paul de Kock, and felt that most of my life had been squandered. But it did not take long before I recognized that it was the best thing I could have done. Within a few weeks, I had won Edison's confidence.*

Nikola Tesla, on meeting Thomas Edison.[1]

When Nikola Tesla arrived in New York on June 6, 1884, he said that meeting Edison was one of the most memorable events in his life. Tesla had already worked for Edison in the French Edison Company outside Paris. The company was introducing Edison's incandescent lighting system to the "City of Light." Tesla had been hired by Tivadar Puskás, whom he had previously worked for in Budapest, installing the first telephone exchange there. In his spare time, Tesla worked on a revolutionary new type of motor that used alternating current, or AC.[2]

"It was the simplest motor I could conceive of," said Tesla. "It had only one circuit, and no windings on the armature or the fields. It was of marvelous simplicity." [3]

No one in the Edison company was interested, largely because Edison had firmly nailed his flag to the mast of DC (direct current). However, Tesla did catch the eye of Charles Batchelor who had been head of the Edison organization in Paris, but he was returning to New York to head the Edison

Nikola Tesla, aged 28, in 1884.

Machine Works there and asked Tesla to come with him. To smooth Tesla's passage into the Edison organization, Batchelor got a letter of introduction from Tivadar Puskás addressed to Edison, saying: "I know two great men and you are one of them; the other is this young man." [4]

FIXING THE SS *OREGON*

In New York, Edison dubbed the slender, elegantly dressed Tesla, who expressed himself in a flowery style in heavily accented English, as "our Parisian." [5]

Tesla proved to be the answer to Edison's prayers. The dynamos the Edison organization had installed to power the lights on the SS *Oregon*, then the holder of the Blue Riband for the fastest transatlantic passenger crossing, had broken down, delaying her sailing. Tesla immediately volunteered to make repairs. He and his team worked overnight and the *Oregon* set sail the following day for another record-breaking run. Tesla recalled the event clearly:

> *The predicament was a serious one, and Edison was much annoyed. In the evening, I took the necessary instruments with me and went aboard the vessel, where I stayed for the night. The dynamos were in bad condition, having several short-circuits and breaks, but with the assistance of the crew, I succeeded in putting them in good shape. At five o'clock in the morning, when passing along Fifth Avenue on my way to the shop, I met Edison with Batchelor and a few others, as they were returning home to retire. "Here is our Parisian running around at night," he said. When I told him I was coming from the Oregon and had repaired both machines, he looked at me in silence and walked away without another word. But when he had gone some distance I heard him remark, "Batchelor, this is a damn good man."* [6]

The dynamo room aboard the SS *Oregon*, after Tesla's repair work.

EDISON ON TESLA

It seems that Edison remembered Tesla from Paris, before he immigrated. Edison told the *Buffalo New York News* in 1896: "Oh, he's a great talker, and, say, he's a great eater too. I remember the first time I saw him. We were doing some experimenting in a little place outside Paris, and one day a long, lanky lad came in and said he wanted a job. We put him to work thinking he would soon tire of his new occupation for we were putting in twenty to twenty-four hours a day then, but he stuck right to it and after things eased up one of my men said to him: 'Well, Tesla, you've worked pretty hard, now I'm going to take you into Paris and give you a splendid supper.' So he took him to the most expensive café in Paris – a place where they broil an extra thick steak between two thin steaks. Tesla stowed away one of those big fellows without any trouble and my man said to him: 'Anything else, my boy? I'm standing treat.' 'Well, if you don't mind, sir,' said my apprentice, 'I'll try another steak.' After he left me he went into other lines and has accomplished quite a little." [7]

TESLA ON EDISON

Tesla told *The New York Times* on October 19, 1931: "I came from Paris in the spring of 1884, and was brought into intimate contact with him. We experimented day and night, holidays not excepted. His existence was made up of alternate periods of work and sleep in the laboratory. He had no hobby, cared for no sport or amusement of any kind and lived in utter disregard of the most elementary rules of hygiene. There can be no doubt that, if he had not married a woman of exceptional intelligence, who made it the one object of her life to preserve him, he would have died many years ago from the consequences of sheer neglect. So great and uncontrollable was his passion for work."

"If he had a needle to find in a haystack he would not stop to reason where it was most likely to be, but would proceed at once with the feverish diligence of a bee, to examine straw after straw until he found the object of his search. His method was inefficient in the extreme, for an immense ground had to be covered to get anything at all unless blind chance intervened, and at first I was almost a sorry witness of his doings, knowing that just a little theory and calculation would have saved him ninety per cent of the labor. But he had a veritable contempt for book learning and mathematical knowledge, trusting himself entirely to his inventor's instinct and practical American sense. In view of this, the truly prodigious amount of his actual accomplishments is little short of a miracle." [8]

WORKING ALL THE HOURS

Edison was famous for working eighteen-hour days and when Tesla went to work at the Edison Machine Works on Goerck Street on the lower east side of Manhattan, he sought to emulate his boss.

"For nearly a year, my regular hours were from 10.30 a.m. to five o'clock the next morning without a day's exception,"[9] Tesla said.

Edison told him: "I have had many hard-working assistants, but you take the cake."[10]

Possibly thinking that Tesla came from Transylvania, Edison asked whether Tesla had ever tasted human flesh. Tesla was also appalled at Edison's "utter disregard of the most elementary rules of hygiene."

Before leaving Paris, Tesla had made a study of unpurified water and shunned it. Later he wrote: "If you would watch only for a few minutes the horrible creatures, hairy and ugly beyond anything you can conceive, tearing each other up with the juices diffusing throughout the water – you would never drink a drop of unboiled or unsterilized water."

Hygiene became a life-long obsession for Tesla. He avoided unsavory restaurants, and examined the crockery and cutlery before eating. He also enquired about Edison's diet.

"You mean to make me so all-fired smart?" Edison replied.

Tesla nodded.

Edison said, perhaps in jest, that he ate a daily regimen of Welsh Rarebit as "it's the only breakfast guaranteed to renew one's mental faculties after the long vigils of toil."[11] Tesla began to do the same despite a protesting stomach. Otherwise, Tesla still took the time to enjoy a good meal – the *table d'hôte* at a restaurant in Greenwich Village with a bottle of red wine – and play billiards, a game he had mastered as a student. According to Edison's personal secretary Alfred O. Tate: "He played a beautiful game. He was not a high scorer but his cushion shots displayed a skill equal to that of a professional exponent of this art."[12]

THE BACKWOODS BOY

Tesla, the cultured European, could not have been more different from Thomas Alva Edison, the boy from the backwoods. Born in Milan, Ohio, in 1847, one of his earliest recollections was seeing the prairie-schooner covered wagons assembling before setting off for California.[13]

Edison had a retentive memory, learning all the songs of the local lumber gangs and canal men before he was five years old. He learned to write by copying laboriously the signs of the village stores. And he was inquisitive. One day, his father found him sitting in a nest full of goose and hens' eggs in the barn, trying to hatch them out.[14]

It was a life not without its hazards. He went swimming in the creek with the son of the wealthiest man in town. The other boy went missing and was found drowned. Edison himself almost drowned when he fell in the Milan Canal. He was almost smothered when he fell in a grain elevator. He lost the top of a finger in an accident with an axe and received a public whipping after burning down a barn.[15]

The family moved to Port Huron, Michigan. Edison's mother was a teacher and by the age of twelve he had read Gibbon's *Decline and Fall of the Roman Empire*, Hume's *History of England*, Sears' *History of the World*, Burton's *Anatomy of Melancholy*, and the *Dictionary of Sciences*. He had even attempted to struggle through Newton's *Principia*, whose mathematics were decidedly beyond both teacher and student. Besides, Edison, like Faraday, was never a mathematician, and had little personal use for arithmetic beyond that which is called "mental." He said once to a friend: "I can always hire some mathematicians, but they can't hire me."[16]

His father, by the way, always encouraged his literary tastes, and paid him a small sum for each new book mastered. Edison also enjoyed novels, particularly the work of Victor Hugo and was nicknamed "Victor Hugo Edison."[17]

THE TRAVELING LABORATORY

At the age of ten or eleven, Edison got a copy of Parker's *School of Natural Philosophy*, an elementary book on physics, and developed a great passion for chemistry, building a laboratory in the cellar of the house.[18]

To pay for the chemicals, he applied to become a newsboy, selling papers on the Grand Trunk Railroad that ran between Port Huron and Detroit. This had the added advantage of giving him access to other newspapers and magazines free of charge, and the opportunity to spend long hours in the public library in Detroit. He made more money taking vegetables on the train to sell in the city, handing his mother over $600 one year.[19]

To occupy his time on the journey, he installed his chemical laboratory in the baggage car with wooden racks made by George Pullman (1831 – 1897) who then had a workshop in Detroit and was already working on his sleeping car. He also acquired a small printing press and began producing, and selling, his own weekly newspaper. *The Times* of London noted that it was the first newspaper in the world to be printed on a train in motion. It made him another $20 to $30 a month.[20]

With the Civil War raging, the demand for newspapers soared. Edison bribed the telegraph operator to wire on ahead news of battle. When the train pulled in, he would quickly sell out.[21]

EDISON'S DEAFNESS

When Edison was thrown off the Grand Trunk Railroad for setting the baggage car on fire, the conductor also boxed his ears, leaving him partly deaf. Edison explained how he coped:

> *This deafness has been of great advantage to me in various ways. When in a telegraph office, I could only hear the instrument directly on the table at which I sat, and unlike the other operators, I was not bothered by the other instruments. Again, in experimenting on the telephone, I had to improve the transmitter so I could hear it. This made the telephone commercial, as the magneto telephone receiver of Bell was too weak to be used as a transmitter commercially. It was the same with the phonograph. The great defect of that instrument was the rendering of the overtones in music, and the hissing consonants in speech. I worked over one year, twenty hours a day, Sundays and all, to get the word 'specie' perfectly recorded and reproduced on the phonograph. When this was done I knew that everything else could be done – which was a fact. Again, my nerves have been preserved intact. Broadway is as quiet to me as a country village is to a person with normal hearing.*[22]

Thomas Edison, aged 10.

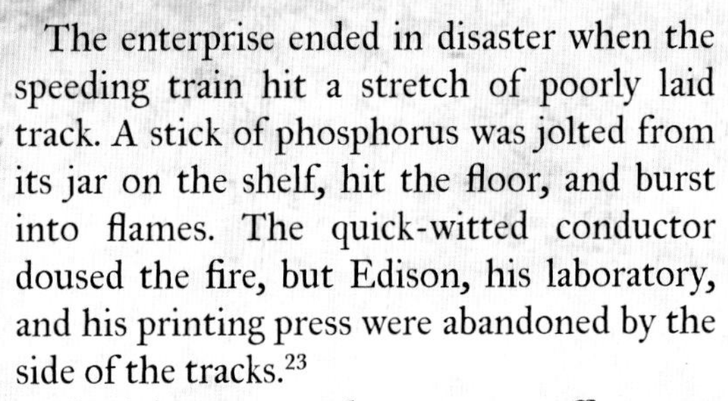

The enterprise ended in disaster when the speeding train hit a stretch of poorly laid track. A stick of phosphorus was jolted from its jar on the shelf, hit the floor, and burst into flames. The quick-witted conductor doused the fire, but Edison, his laboratory, and his printing press were abandoned by the side of the tracks.[23]

The laboratory and newspaper office were set up again at home. Being easily accessible in one place had its disadvantages. When one target of his pen took offence, Edison found himself hurled in the St. Clair River.[24]

ROAMING TELEGRAPH OPERATOR

The Grand Trunk Railroad had machine shops in Port Huron, where Edison became thoroughly familiar with the intricacies of fire-box, boiler, valves, levers, gears, and the like. Then, when visiting friends in the telegraph office, he got interested in electricity. His knowledge of chemistry had already made him familiar with batteries and he experimented with static electricity on cats, making their hair stand on end.

With a friend, he set up a private telegraph system, using stove-pipe wire and bottles set on nails driven into trees as insulators. He made the keys himself in a gun-shop in Detroit and learned Morse code. With the Civil War still going on, he found himself employed by the US Military Telegraph Corps.[25]

This led to a job on the railroad where a senior colleague would leave him copies of *Scientific American* to read. He went on to roam the Midwest as a telegraph operator. He was in the South during Reconstruction and worked on ways to improve the telegraphic "repeater" that boosted the signal over long distances.

During the Civil War, telegraph messages had been intercepted and Edison began work on his quadruplex system, sending four messages down the same line, rendering each single message undecipherable.[26]

14-year-old Edison prints the first edition of the *Grand Trunk Herald* on board a moving train in 1862.

WORKING FOR WESTERN UNION

Edison went to work for Western Union in Boston, operating the New York number one wire. He often visited a workshop where telegraph equipment was repaired. There, with others, he came up with an instrument that earned him his first patent, No. 90,646, which was taken out on June 1, 1869. It was a device designed to count the votes in the House of Representatives, taking only a minute or so.[27]

Among the other telegraph operators in Boston were young men studying at Harvard. But when it came to practical things, Edison easily outstripped them. He installed a device made of two sheets of tinfoil connected to a battery. When a cockroach crawled across them, it was vaporized in a blinding flash. This was a version of an apparatus that he had used to paralyze rats out West.[28]

"After the vote-recorder I invented a stock ticker, and started a ticker service in Boston; had thirty or forty subscribers, and operated from a room over the Gold Exchange,"[29] said Edison. Other stock ticker systems were already in use in New York.

AUTOMATING THE TELEGRAPH

After borrowing $800, he developed a duplex system, so two messages could be sent down a single telegraph wire at the same time, doubling the traffic. This took him to New York where he visited the Gold Indicator Company. Edison said:

> *While sitting in the office, the complicated general instrument for sending on all the lines, and which made a very great noise, suddenly came to a stop with a crash. Within*

ASSASSINATION OF PRESIDENT LINCOLN

Edison was the telegraph operator in Cincinnati on the night of Friday, April 14, 1865, when Lincoln was shot. He said:

> *I noticed an immense crowd gathering in the street outside a newspaper office. I called the attention of the other operators to the crowd, and we sent a messenger boy to find the cause of the excitement. He returned in a few minutes and shouted "Lincoln's shot." Instinctively the operators looked from one face to another to see which man had received the news. All the faces were blank, and every man said he had not taken a word about the shooting. "Look over your files," said the boss to the men handling the press stuff. For a few moments we waited in suspense, and then a man held up a sheet of paper containing a short account of the shooting of the President. The operator had worked so mechanically that he had handled the news without the slightest knowledge of its significance.*[30]

Thomas Edison, the Telegraph Operator, aged 16.

two minutes over three hundred boys – a boy from every broker in the street – rushed up-stairs and crowded the long aisle and office, that hardly had room for one hundred, all yelling that such and such a broker's wire was out of order and to fix it at once. It was pandemonium ... I went to the indicator, and, having studied it thoroughly, knew where the trouble ought to be, and found it.[31]

Edison fixed it and was instantly hired at a salary of $300 a month. He set to work on an improved stock ticker. When he demonstrated it, the president of the Gold & Stock Telegraph Company gave him $40,000 for it. With the money, he bought machinery and set up a business in Newark, New Jersey, manufacturing his new improved stock tickers.[32]

He then went to work on an automatic telegraph that sent and recorded a thousand words a minute between New York and Washington, and sent 3,500 words a minute to Philadelphia – while a skilled operator with a manual key could manage no more than forty or fifty words a minute. Edison's machine printed out the results, not as dots and dashes, but in letters using "typewriters."[33]

Edison went to England and set up a line between London and Liverpool, again transmitting a thousand words a minute. He also experimented with submarine cables.

Returning to the US, he set about perfecting duplex and quadruplex telegraphy, racking up patents along the way. By 1910, it was estimated that his quadruplex had saved between $15 million and $20 million in line construction in America alone.

DEVELOPING TELECOMMUNICATIONS

When Alexander Graham Bell (1847 – 1922) invented the telephone in 1876, he described the results he obtained with his first apparatus as "unsatisfactory and discouraging." The president of Western Union, William Orton, set Edison to the task of improving it.[34] Edison said:

Bell invented the first telephone, which consisted of the present receiver, used both as a transmitter and a receiver – the magneto type. It was attempted to introduce it commercially, but it failed on account of its faintness and the extraneous sounds which came in on its wires from various causes. Mr. Orton wanted me to take hold of it and make it commercial. As I had also been working on a telegraph system employing tuning-forks, simultaneously with both Bell and Gray, I was pretty familiar with the subject. I started in, and soon produced the carbon transmitter, which is now universally used.[35]

Edison also devised a chalk earpiece called a motograph to get round Bell's patent on the electromagnetic receiver. In this device, the user turned the handle which caused a diaphragm in the receiver to be tensioned by the friction created between a platinum strip and a revolving drum coated with damp chalk. Varying electrical currents from the

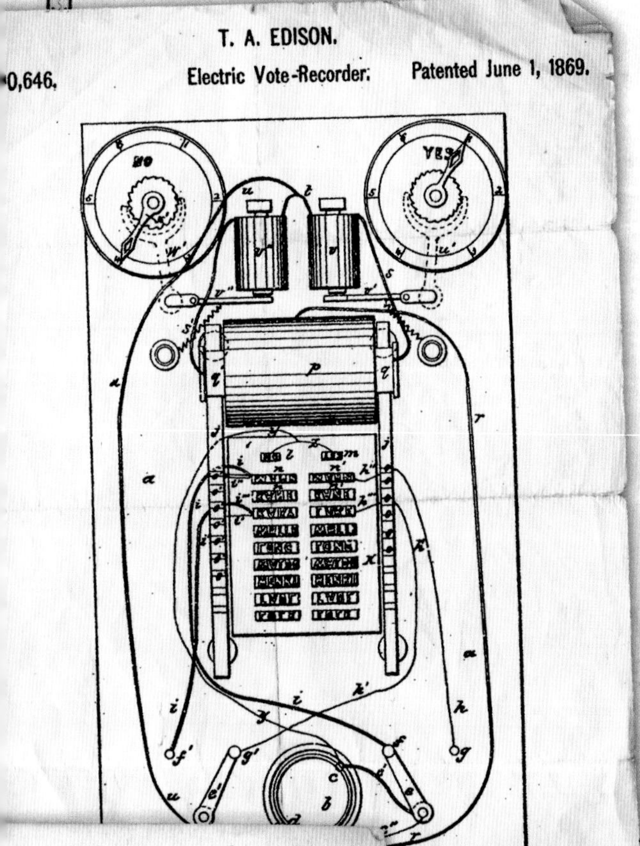

Edison's patent for the Electric Vote-Recorder, June 1, 1869.

transmitter controlled the friction and the tension, making the diaphragm produce sound.[36]

Bell's receiver proved superior and the telephone system developed with Edison's mouthpiece and Bell's earpiece.[37]

ACOUSTIC TELEGRAPHY AT MENLO PARK

In 1876, Edison had enough money to set up Menlo Park, the world's first industrial research lab in Middlesex County, New Jersey.[38] He worked there with two key associates – Swiss-born machinist John Kruesi and Charles Batchelor, the English master mechanic who first brought Tesla to Edison's attention.[39]

At the time, the telephone was not yet seen as a means of person-to-person communication. Rather it was considered "acoustic telegraphy" and, as Edison had developed automatic telegraphy where messages were mechanically transcribed, at Menlo Park he set about doing the same thing for the telephone. The result was the phonograph.[40]

To merchandize his new invention, Edison took offices at 203 Broadway in New York. The tin-foil cylinder was soon replaced by a wax cylinder that could then be erased and used again and again. Edison also realized that another commercial use for his invention was the duplication of recordings – thousands of copies could be made from one "master." He set up the National Phonograph Company to exploit the phonograph for musical purposes as a competitor to the music-box. To do so, he developed a simple motor with a governor, maintaining the rate of rotation accurately enough to reproduce music.[41]

EDISON'S MARRIAGE AND CHILDREN

Unlike the ascetic Tesla, Edison married twice. In 1871, he married 16-year-old Mary Stilwell, who gave him three children – the first two were nicknamed "Dot" and "Dash." Mary died in 1884, aged 29, of unknown causes.

Two years later, the 39-year-old Edison married 20-year-old Mina Miller. He moved out of Menlo Park and bought *Glenmont* in West Orange, where he built new laboratories. They also had a winter retreat in Fort Myers, Florida.

Mina also gave him three children and outlived him, dying in 1947. The two sons from his first marriage were both reprimanded by their father for selling the use of the Edison name to endorse dubious inventions. The two sons from his second marriage worked for their father's corporation.[42]

Mina and Thomas Edison, 1886.

INVENTING THE MICROPHONE

Alexander Graham Bell and Elisha Gray (1835 – 1901) used a liquid transmitter in their early experiments. A wire attached to the bottom of a parchment diaphragm was adjusted so that it just barely made contact with the water, which was made electrically conductive with a small amount of acid. Words spoken above the diaphragm cause it to flex up and down, making the attached wire have more or less contact with the acidulated water, thereby changing the circuit resistance. The resulting current variations in the listening device reproduce the original sounds.

Bell then came up with a microphone where the diaphragm was attached to the armature of electromagnet. Movements of the diaphragm then induced an electrical signal in the coil of the electromagnet. These were weak, so could not be transmitted very far.

The carbon-button microphone had two electrical contacts separated by a thin layer of carbon which is compressed by the flexing of the diaphragm. This alters the electrical resistance of the carbon and, thus, the current in the circuit. The strength of the signal depends not on weak electromagnetic induction, but on potential of the battery.[43]

The carbon-button microphone was patented by Emile Berliner (1851 – 1929) in 1877. Bell was so impressed that he bought the rights from Berliner for $50,000. However, in 1892, the US Supreme Court found in favor of Thomas Edison.[44]

In the Edison system, he also incorporated a transformer, so that the higher current in the microphone circuit could be turned into a higher voltage that could be transmitted hundreds of miles without significant loss.[45]

Berliner got his own back with his invention of the gramophone, a disk-version phonograph that overtook Edison's cylinder device.[46]

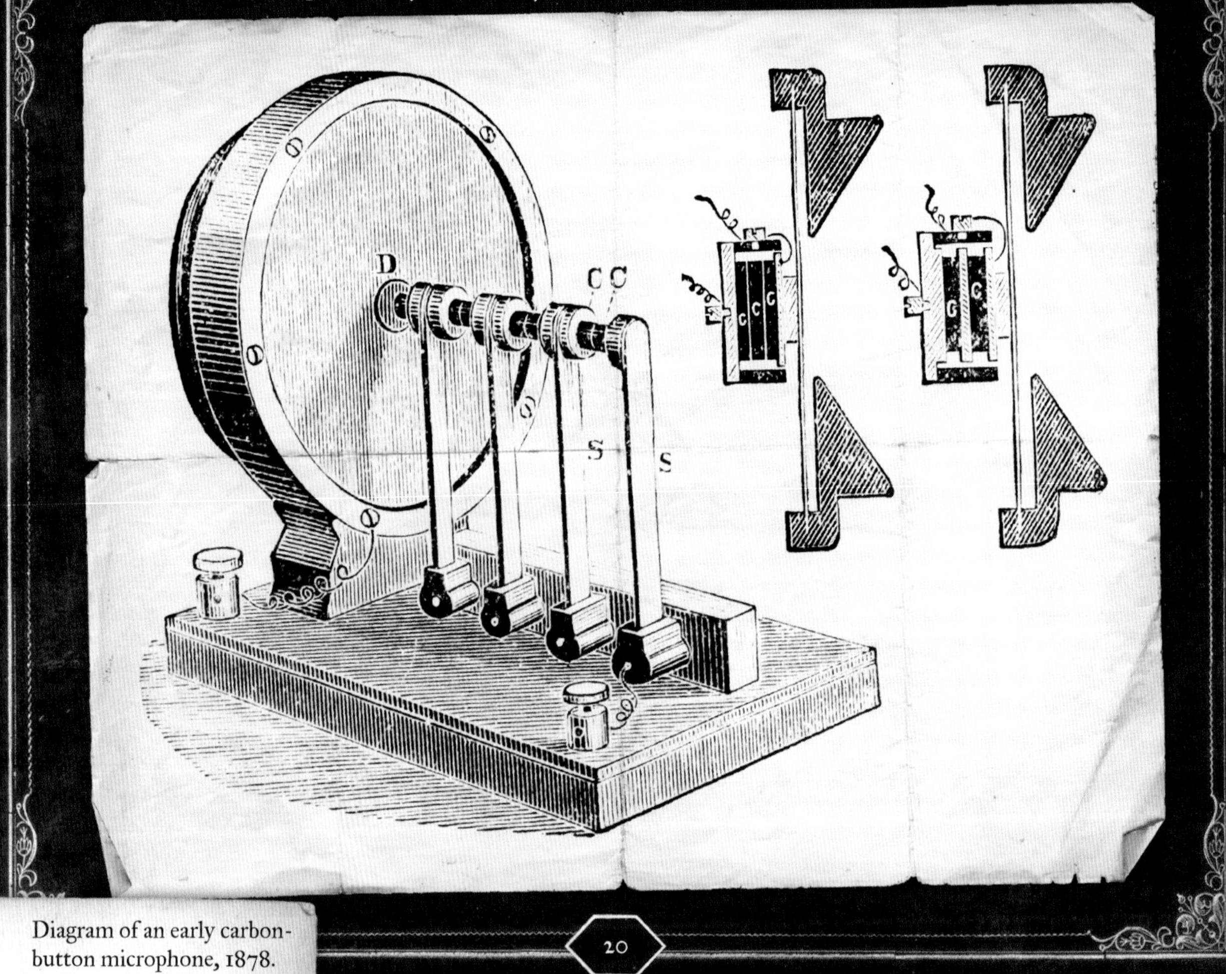

Diagram of an early carbon-button microphone, 1878.

THE INVENTION OF THE TELEPHONE

On February 14, 1876, Alexander Graham Bell filed a patent application for his telephone. Two hours later, electrical engineer Elisha Gray filed a caveat for *his* telephone. The inventions described in the documents were almost exactly the same. In the months that followed, there was speculation that a crooked patent office clerk had reversed the order in which the documents were filed, but in the resulting court case, Bell was awarded the patent outright.[47]

However, among the papers found at Edison's West Orange Laboratory was a lithograph, the size of an ordinary patent drawing, headed "First Telephone on Record." The device shown, made by Edison in 1875, was included in a caveat filed January 14, 1876, a month before Bell or Gray. It showed a little solenoid arrangement, with one end of the plunger attached to the diaphragm of a speaking or resonating chamber. Edison said that, while the device is crudely capable of use as a telephone, he did not invent it for transmitting speech, but as an apparatus for analyzing the complex waves arising from various sounds used in his investigations of harmonic telegraphs. He did not try it with the human voice until Bell filed his patent. Only then did he discover that it could be used as a telephone. Nevertheless Edison always gave Bell credit for the discovery of the transmission of articulate speech by talking against a diaphragm placed in front of an electromagnet.[48]

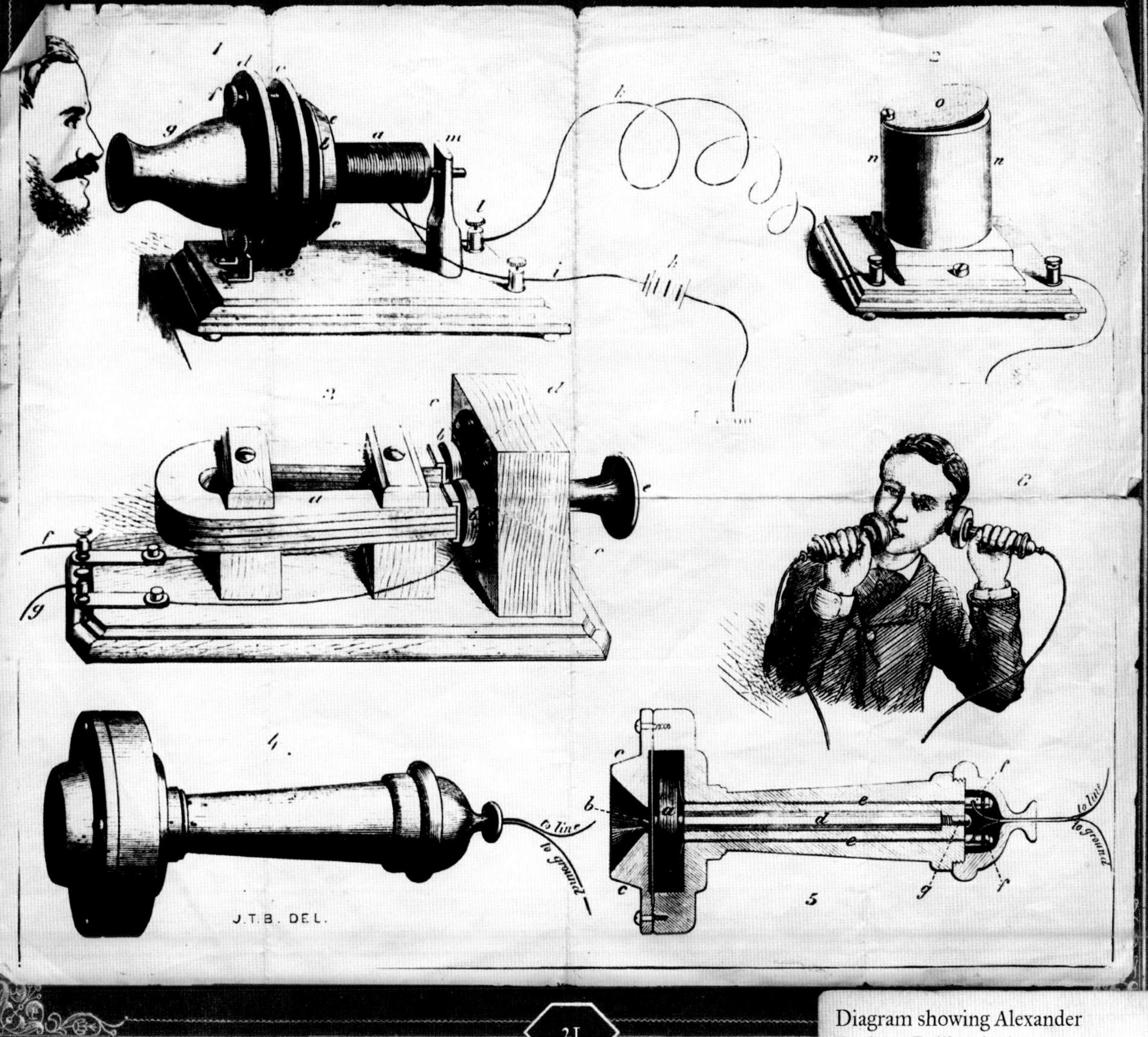

Diagram showing Alexander Graham Bell's telephone system.

EDISON'S PHONOGRAPH

Edison began by experimenting with an automatic method of recording telegraph messages on a disk of paper laid on a revolving platen with a spiral groove on its surface. An electromagnet with the embossing point connected to an arm traveled over the disk and any signal passing through the magnets were embossed on the disk of paper.

"From my experiments on the telephone I knew of the power of a diaphragm to take up sound vibrations, as I had made a little toy which, when you recited loudly in the funnel, would work a pawl connected to the diaphragm; and this engaging a ratchet-wheel served to give continuous rotation to a pulley," he said. "This pulley was connected by a cord to a little paper toy representing a man sawing wood. Hence, if one shouted: 'Mary had a little lamb,' etc., the paper man would start sawing wood. I reached the conclusion that if I could record the movements of the diaphragm properly, I could cause such records to reproduce the original movements imparted to the diaphragm by the voice, and thus succeed in recording and reproducing the human voice."

Thomas Edison and his early phonograph, April 1878.

"Instead of using a disk I designed a little machine using a cylinder provided with grooves around the surface. Over this was to be placed tinfoil, which easily received and recorded the movements of the diaphragm ... The workman who got the sketch was John Kruesi. I didn't have much faith that it would work, expecting that I might possibly hear a word or so that would give hope of a future for the idea. Kruesi, when he had nearly finished it, asked what it was for. I told him I was going to record talking, and then have the machine talk back. He thought it absurd. However, it was finished, the foil was put on; I then shouted 'Mary had a little lamb,' etc. I adjusted the reproducer, and the machine reproduced it perfectly. I was never so taken aback in my life. Everybody was astonished."[49]

Kruesi exclaimed: "*Mein Gott im Himmel!*"[50] Frederick Converse Beach, the editor of *Scientific American* would be even more impressed.

"That morning I took it over to New York and walked into the office of the *Scientific American*, went up to Mr. Beach's desk, and said I had something to show him," said Edison. "He asked what it was. I told him I had a machine that would record and reproduce the human voice. I opened the package, set up the machine and recited, 'Mary had a little lamb,' etc. Then I reproduced it so that it could be heard all over the room. They kept me at it until the crowd got so great Mr. Beach was afraid the floor would collapse; and we were compelled to stop. The papers next morning contained columns. None of the writers seemed to understand how it was done. I tried to explain, it was so very simple, but the results were so surprising they made up their minds probably that they never would understand it – and they didn't. I started immediately making several larger and better machines, which I exhibited at Menlo Park to crowds. The Pennsylvania Railroad ran special trains."[51]

Newspapers worldwide carried the story. In Paris, *Le Figaro* got carried away. In their account, under the headline "This Astounding Edison," they added a fanciful description of Edison's new "aerophone," a steam machine that carried the voice a distance of one and a half miles. "You speak to a jet of vapor," it said. "A friend previously advised can answer you by the same method."[52]

THE FUTURE OF THE PHONOGRAPH

Writing in 1878 in the *North American-Review*, Edison listed his own ideas as to the future applications of the new invention:

1. Letter writing and all kinds of dictation without the aid of a stenographer.
2. Phonographic books, which will speak to blind people without effort on their part.
3. The teaching of elocution.
4. Reproduction of music.
5. The 'Family Record' – a registry of sayings, reminiscences, etc., by members of a family in their own voices, and of the last words of dying persons.
6. Music-boxes and toys.
7. Clocks that should announce in articulate speech the time for going home, going to meals, etc.
8. The preservation of languages by exact reproduction of the manner of pronouncing.
9. Educational purposes; such as preserving the explanations made by a teacher, so that the pupil can refer to them at any moment, and spelling or other lessons placed upon the phonograph for convenience in committing to memory.
10. Connection with the telephone, so as to make that instrument an auxiliary in the transmission of permanent and invaluable records, instead of being the recipient of momentary and fleeting communication.[53]

THIS ASTOUNDING EDISON

For the eclipse of the Sun in 1878, Edison developed a microtasimeter that could measure infrared radiation down to a millionth of a degree Fahrenheit. While out in Wyoming with a party of astronomers, he noticed the long distance that farmers had to transport their grain and proposed the construction of a network of light electric railways that would be operated automatically.

Electric arc lights were used at the time. These produced an intense white light and Edison speculated that, with a device like his microtasimeter, he could subdivide an arc to produce smaller domestic lights analogous to individual gas burners.[54]

As early as 1845, inventors had noticed that light was given off when electricity, passing through a wire, heated it. However, by 1877, leading electricians, physicists, and experts of the period had been studying the subject for more than a quarter of a century, and had proven mathematically and by close reasoning that the "subdivision of the electric light," as it was then termed, could not be done.[55]

"Much nonsense has been talked in relation to this subject. Some inventors have claimed the power to 'indefinitely divide' the electric current, not knowing or forgetting that such a statement is incompatible with the well-proven law of conservation of energy,"[56] said Dr. Paget Higgs, possibly referring to Edison.

At a lecture at the Royal United Service Institution, leading English scientist, William Preece (1834 – 1913), who introduced the telephone to Great Britain and later sponsored the work of Guglielmo Marconi, said: "The subdivision of the light is an absolute *ignis fatuus*[57] ... a will o' the wisp." Such opinions always proved a stimulus to Edison.

THE EDISON ELECTRIC LIGHT COMPANY

J.P. Morgan and the Vanderbilts set up the Edison Electric Light Company and advanced Edison $30,000 for research and development. Dogged by failed experiments, Edison turned his attention to electrical generation. The resulting generator, by reverse action, also worked as a motor.[58]

One of Edison's staff came up with an incandescent light with a platinum filament inside an evacuated glass bulb. But the cost of platinum made this impractical. The English physicist Joseph Swan had succeeded in making a bulb with a carbon filament. Edison set to work on this and on November 4, 1879,

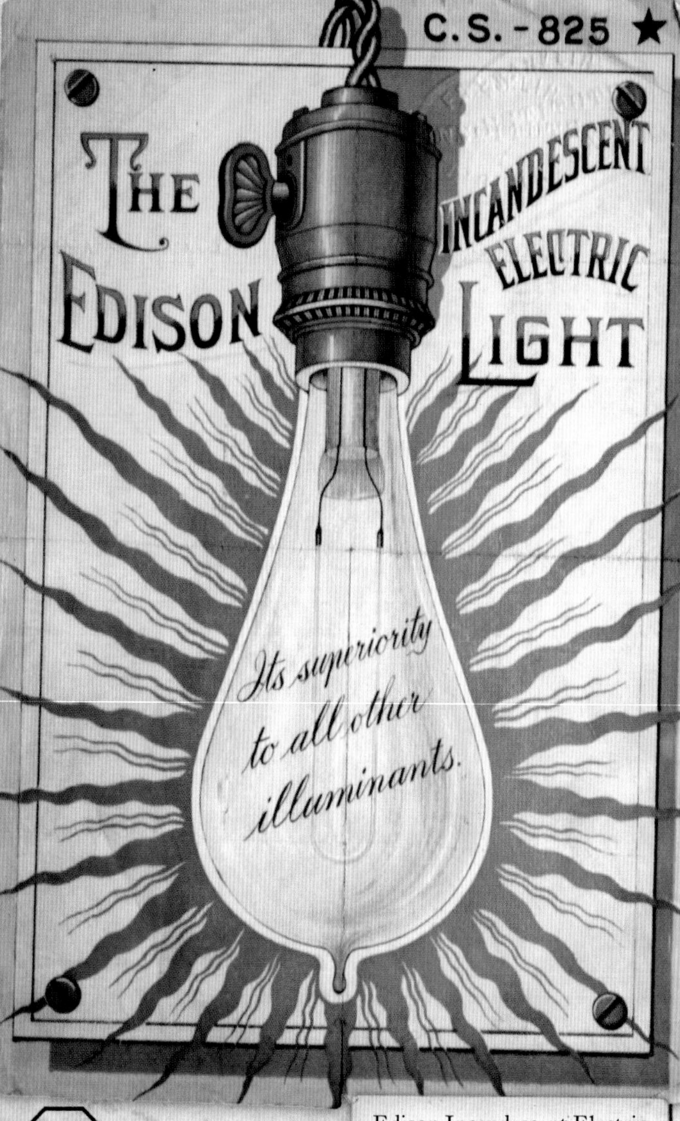

Edison Incandescent Electric Light catalog, 1887.

filed a US patent on the first commercially practical electric light.

Edison and his team soon found that carbonized bamboo made the best filament. Demonstrating his invention at Menlo Park, he said: "We will make electricity so cheap that only the rich will burn candles."[59]

He installed the world's first electric lighting system on the steamship *Columbia* in 1880. Then came the printing firm of Hinds, Ketcham & Company in New York,[60] followed by lighting the Holborn Viaduct in London.[61]

One of Edison's biggest critics, the physicist William Preece, was then forced to reassess his view on the impossibility of subdividing light:

> *Mr. Edison's system has been worked out in detail, with a thoroughness and mastery of the subject that can extract nothing but eulogy from his bitterest opponents ... Many unkind things have been said of Mr. Edison and his promises; perhaps no one has been severer in this direction than myself. It is some gratification for me to be able to announce my belief that he has at last solved the problem he set himself to solve.*

PEARL STREET POWER STATION

Edison built his first power station on Pearl Street in lower Manhattan in 1882, initially to power 400 lamps owned by 85 customers. By 1884, it was serving 508 customers with 10,164 lamps.[62] The electricity was carried above ground on poles with dozens of crooked crossbeams supporting sagging wires. The exposed electrical wiring was a constant danger and unsuspecting children climbing the poles would suffer lethal electric shocks. In spite of the perils, wealthy New Yorkers rushed to have their homes wired, the most important being Edison's backer, J.P. Morgan.[63]

Steam engines were used to power generators, producing 110 volts of direct current.[64] This comparatively low voltage meant that the power station could only supply the local area and the fact that it was direct current meant that it could not easily be converted to a higher voltage for long distance transmission. The young Nikola Tesla who had just come into Edison's employ was about to change all that.[65]

The Dynamo Room, Pearl Street Power Station.

THE INCANDESCENT LAMP

In 1802, Humphry Davy (1778 – 1829) ran a current through a platinum wire, heating it until it became incandescent, but he did not pursue this method of producing light. In 1835, Scottish inventor James Bowman Lindsay (1799 – 1862) demonstrated a similar device at a public meeting in Dundee, Scotland, stating he could "read a book at a distance of one and a half feet."[66] His diverse interest took him on to wireless telegraphy and astronomy but his innovations in various fields were not exploited until long after his death.

Five years later, British scientist Warren de la Rue (1815 – 89) passed an electric current through a coiled platinum filament in a vacuum tube, but the cost of platinum made this impractical.

Then in 1845, American inventor John W. Starr (1822 – 46) took out an English patent for a "continuous metallic or carbon conductor intensely heated by the passage of electricity for the purpose of illumination."[67] He proposed enclosing a heated carbon rod in a vacuum tube, but he died the following year. Other inventors in France, Germany, and Russia were working along the same lines.

English inventor Joseph Wilson Swan (1828 – 1914) demonstrated a working device in 1860, but pumps then could not make a good-enough vacuum. Then they improved. Swan returned to his investigations and on December 18, 1878, demonstrated a light bulb using a thin carbon rod.

Meanwhile Edison was experimenting with various materials. On October 21, 1879, he produced a bulb with a filament made of a carbonized cotton sewing thread. It gave off light steadily for two days. Edison patented his carbon filament, but lost a patent infringement suit against Swan in 1882. The two men reached a compromise. They set up the Edison and Swan United Electric Light Company Limited the following year, leading to the successful international marketing of the incandescent lamp.

The carbon filament was replaced by tungsten in the early years of the twentieth century.

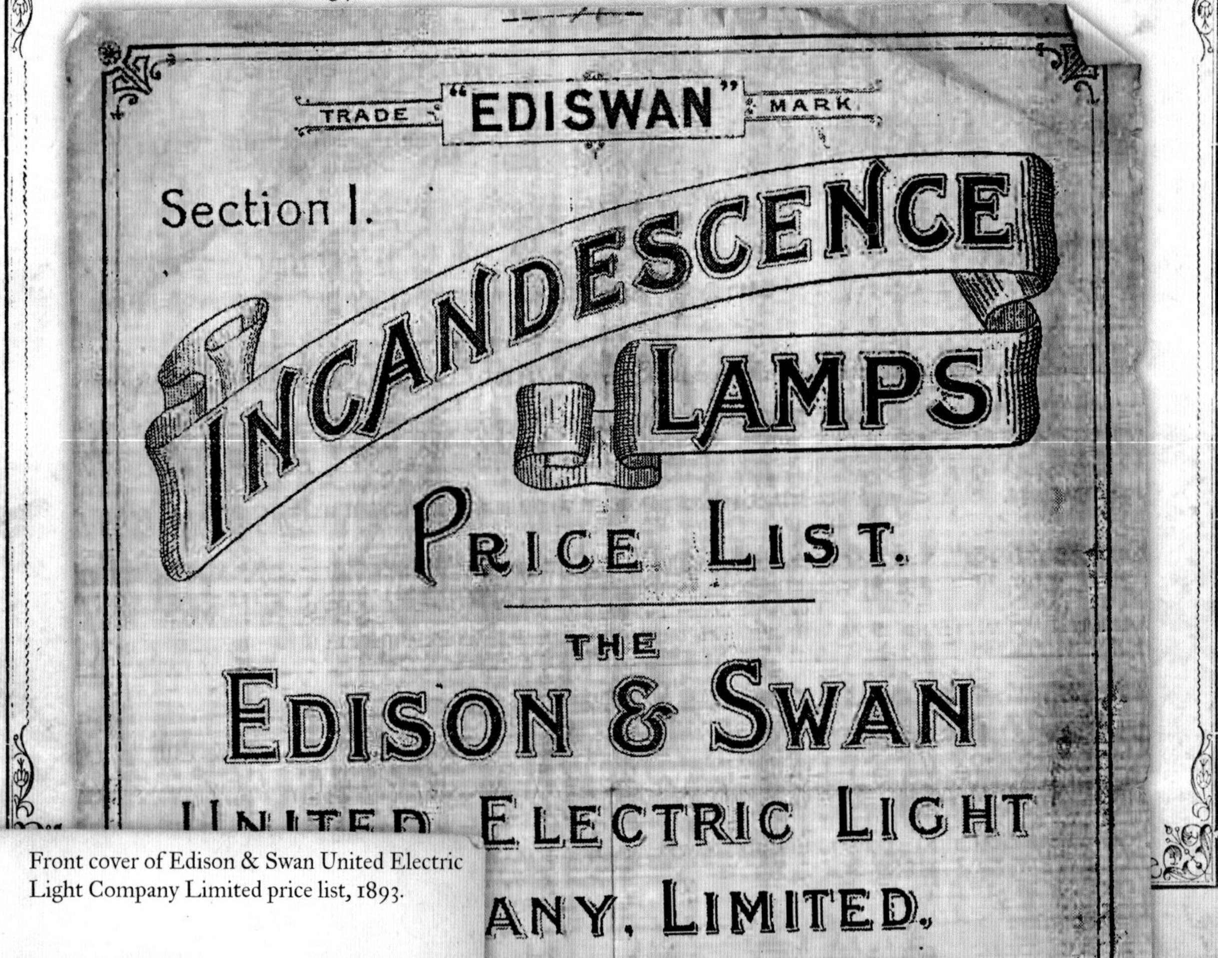

Front cover of Edison & Swan United Electric Light Company Limited price list, 1893.

SIR JOSEPH WILSON SWAN
(1828 – 1914)

Apprenticed to a druggist in his hometown of Sunderland, north-east England, Swan went on to became a partner in a chemical firm.

He took out his first patent in 1864. It was for the "carbon process" for printing photographs with permanent pigments. His firm began to make dry plates and he patented the first bromide paper in 1879.

From 1848 he had also been experimenting with electric light. In 1860, he developed a primitive electric lamp, using a filament of carbonized paper in an evacuated, stoppered glass bulb. But the lack of a good vacuum and an adequate source of electricity gave it a short lifetime and a dim light.

In 1877, using the better vacuum pump then available, he made a bulb with a carbon filament treated in sulfuric acid inside a hermetically sealed glass bulb. He formed a company to make his lamps, and amazed Londoners in 1881, lighting up the Savoy Theater with 1,200 of them.

His other inventions include the chrome tanning of leather and the cellular lead plate for rechargeable batteries, patented in 1881. He was knighted in 1904.

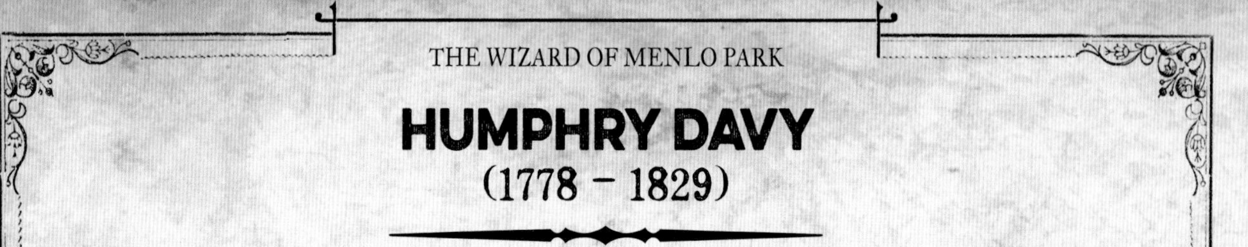

HUMPHRY DAVY
(1778 – 1829)

Born in Penzance, Cornwall, England, Davy's first love was poetry, but when his father died in 1795, he was apprenticed to a surgeon-apothecary. Befriended by Davies Giddy (later Gilbert, president of the Royal Society 1827 – 30), he began the study of science.

In 1798, he joined the newly founded Medical Pneumatic Institution to investigate the therapeutic properties of various gases. His paper *Researches, Chemical and Philosophical, Chiefly Concerning Nitrous Oxide, or Dephlogisticated Nitrous Air, and its Respiration,* in 1800 established his reputation. The following year he was invited to become a lecturer at the Royal Institution of Great Britain, recently founded in London.

Davy plunged into the investigation of electrolysis and, in 1807, he isolated the metals sodium and potassium using electricity. Further experimentation led to the isolation of boron, chlorine, iodine, and calcium – though it was said that his greatest discovery, in 1812, was Michael Faraday.

In 1815, he invented the miners' safety lamp and became president of the Royal Society in 1820.

JAMES BOWMAN LINDSAY
(1799 – 1862)

In the Scottish *Dundee Advertiser,* April 11, 1834, Lindsay proclaimed that "houses and towns will in a short time be lighted by electricity instead of gas, and heated by it instead of coal; and machinery will be worked by it instead of steam."[68] The following year he claimed to have developed a continuously burning electric light, though few details survive.

In 1845, he sent a proposal to the *Dundee Advertiser* for an autograph telegraph, that used vibrating needles to etch messages into pith balls. When the newspaper carried a suggestion for a submarine Atlantic telegraph cable, Lindsay detailed how a cable might be constructed using copper wire with joints welded by electricity. This was probably the first suggestion of the use of electricity for welding.

He demonstrated the underwater wireless telegraph in 1853, patenting it the following year. Using batteries connected to metal plates submerged in the water, Lindsay carried out several demonstrations in Scotland and England between 1853 and 1860. It worked over short distances, but was unworkable over long distances. To maximize its effectiveness, it needed a line laid on dry land that was longer than the stretch of water the signal was crossing.

Lindsay used astronomy and philology to test the historical accuracy of the Bible, but never completed his *Pentecontaglossal Dictionary*, a comparison of fifty languages begun in 1828.

AC vs DC

Direct current, or DC, is the type of current you get from a battery. It goes in one direction from one terminal to the other. It is what was needed to run a telegraph system. That is why Edison built his Pearl Street Station to produce DC, and he was wedded to DC throughout his career.

Alternating current, AC, goes one way, then the other. In a dynamo or generator, a coil rotates in a magnetic field. This movement produces AC. It can be made to produce DC by the use of a commutator that reverses the connections to the rotating coil. A commutator can also be used to run an electric motor using DC. Without one, a motor will run on AC. However, commutators create sparking and other losses.

The greatest advantage of AC comes from its use in electrical distribution. Because of electrical resistance in wires, a current passing through them produces heat, wasting energy. But if you use electricity at higher voltage – 110,000 volts, say – you only need a thousandth of the current to transmit the same amount of energy at 110 volts and, consequently, only suffer, a thousandth of the heat loss.

Using AC, voltages can be stepped up or down with a transformer. Inside a transformer, there are coils of wire wound around a single iron core. When an electrical current is passed through one of them, it magnetizes the iron core. This, in turn, induces an electric current in the other one. The voltage is stepped up or stepped down according to the ratio of the number of turns of wire in each coil. However, induction only works when the electrical current is being switch on and off again. It only works with an alternating current. DC is no use.

A DC power station, using voltages that can be used around the home, has a limited range before the heat losses grow too great. If AC is used, the voltage supplied by the generator can be stepped up – typically to between 138,000 and 230,000 volts – transmitted long distances, then stepped down again to 110 volts for domestic use.

THE ARC LIGHT

SOUTH FORELAND LIGHTHOUSE, ST. MARGARET'S BAY

At the Royal Institution in London, circa 1809 – 10, Sir Humphry Davy used the current from a battery of two thousand cells to produce an intense voltaic arc between two points of sticks of charcoal which gradually burnt away. For more than thirty years the arc light remained an expensive laboratory experiment, but the coming of the dynamo made arc lighting a commercial prospect. The zinc and acids in the cheapest battery cost twenty times as much as the electricity from a dynamo driven by steam-engine.

From 1850, rapid advances were made in both the dynamo and the arc lamp and, in 1858, under the supervision of Michael Faraday (1791 – 1867), beams of intense arc light were shed over the waters of the Straits of Dover from the beacons of South Foreland and Dungeness. At the Philadelphia Centennial Exposition of 1876, arc lights were used to illuminate the Wallace - Farmer dynamos built at Ansonia, Connecticut, that were powering them.

THE EDISON EFFECT

In 1881, a young engineer named William J. Hammer was testing bulbs and noticed a ghostly blue glow. This effect was nicknamed "Hammer's Phantom Shadow." When Edison's bulbs went on sale in 1883, it became known as the "Edison effect." It was later discovered that this was caused by the thermionic emission of electrons. The phenomenon was put to use in electron tubes, the principal device of the early electronics industry.

AVOIDING THE AC SUBJECT

By the time Nikola Tesla arrived in America, in 1884, Edison was world famous as the inventor of the phonograph and the pioneer of electric light. Born in Croatia in 1856, Tesla, an unknown Serbian, was destined to become Edison's nemesis, but he would always remain an outsider.[1]

Tesla's father was an Orthodox priest and wanted Nikola to follow him into the church. His mother could not read, but Tesla attributed his phenomenal memory to her as well as his capacity for hard work.[2] Both his parents invented implements to help harvest the crops they subsisted on and to relieve the burden of household chores.[3]

Tesla had an interest in electricity from his earliest childhood, after observing the static produced when he stroke the family's cat.[4] He claimed his first invention was a fishing line, noting that frogs would bite onto the hook he had fashioned though it was unbaited. Then he tried to make a primitive helicopter by harnessing four June bugs. However, he always felt he was in the shadow of his older brother Dane who, he said, was "gifted to an extraordinary degree." But Dane was thrown from a horse and died of his injuries when Nikola was five. Tesla said he had nightmares about his brother's death for the rest of his life.[5]

CHILDHOOD TRAUMAS

Like Edison, Tesla had an adventurous and accident-prone childhood. As a baby he had been laid outside in the sunshine, naked, when he was attacked by a goose that seized him by the navel with its beak, and almost pulled it inside out. He once fell headlong into a huge

Smiljan, Croatia. Tesla's birthplace.

vat of boiling milk, risked drowning when swimming under a raft, and found himself almost swept over a waterfall created by a nearby dam. On other occasions, he claimed to have had "hairbreadth escapes from mad dogs, hogs, and other wild animals."[6]

He also had a strange aversion to women's earrings, pearls, peaches, and camphor, and he would not touch other people's hair "except, perhaps, at the point of a revolver." Plainly autistic, he would count his steps as he walked and calculate the volume of soup plates, coffee cups and pieces of food. Every repeated action had to be done a number of times that was divisible by three. If not, he would start over.[7]

Nikola Tesla's father, Milutin Tesla.

FROM ZERO TO HERO

His father took over the church in the town of Gospic, where Tesla started school. Painfully shy, he missed the countryside and found himself ill-equipped for life in the town. To overcome what he called his "weak and vacillating" character, he sought inspiration in the historical novel called *Abafi*. In it, the young roué Olivér Abadir gradually mends his ways and becomes a national hero in Transylvania's fight against the onslaught of the Hungarians, Turks and Austrians. Following Abadir's example, Tesla set about developing willpower.[8]

Tesla had been ostracized by the townsfolk of Gospic after he had accidentally stepped on the train of a grand dame as she left church, ripping it off. He got an opportunity to redeem himself when the town's recently organized fire department was showing off its new fire engine. The entire populace turned out, but when the order to start pumping was given, not a drop of water came out of the hose. While the bewildered firemen examined their shiny new equipment, Tesla realized that there was a blockage in the suction hose that drew water from the river, so he waded in and cleared it. Hero of the day, he found himself carried shoulder high through the streets.[9]

DREAMING OF NIAGARA

At ten, Tesla entered the local Real Gymnasium – the equivalent of an American junior high school. It had a well-equipped physics department which fascinated Tesla:

> *I was interested in electricity almost from the beginning of my educational career. I read all that I could find on the subject ... [and] experimented with batteries and induction coils.*[10]

He was also keen on waterwheels and turbines, and experimented with a flying machine design which, he realized later, could not work because it depended on perpetual motion. Then, after seeing a picture of Niagara Falls, he told his Uncle Josif that one day he would go to America and put a big wheel under the falls to harness its power.[11]

Finishing at the Real Gymnasium at the age of fourteen, Tesla fell ill. While he was recuperating, the local library sent all the books it had not cataloged for him to read and classify.[12]

When he recovered, his father sent him to the Higher Real Gymnasium to prepare him for the seminary, though the prospect of becoming a priest filled Tesla with dread. At his new school, he showed early signs of genius, performing integral calculus in his head, leading his tutors to think he was cheating. In the physics department, he became fascinated by the Crookes radiometer.[13] This consisted of four metal vanes, polished on one side, blackened on the other, mounted on a vertical pivot in a glass bulb. It spun when bright light fell on it. Then, in 1870, for the first time, he saw a steam train.

ELECTRIFIED BY ENGINEERING

After completing his studies, Tesla came down with cholera. He was at death's door, when his father tried to encourage him to rally. Nikola seized the opportunity and said to his father: "Perhaps I may get well if you will let me study engineering."[14]

His father promised that he would go to the best technical institution in the world. Tesla recovered and his father was as good as his word, securing a scholarship for him to study at the Joanneum Polytechnic in Graz, Austria.

The Polytechnic had recently bought a Gramme dynamo which had a commutator so it could turn using the direct current from a battery. The commutator was a split ring on the end of the shaft that was connected to the coil. External connection was made via two brushes rubbing against it. As the commutator turned, it would reverse the connection. Tesla later remembered the scene:

(Left) Young Nikola Tesla studying an early waterwheel experiment. (Right) A Gramme dynamo with split ring commutator.

While Professor Pöschl was making demonstrations, running the machine as a motor, the brushes gave trouble, sparking badly, and I observed that it might be possible to operate a motor without these appliances. But he declared that it could not be done and did me the honor of delivering a lecture on the subject, at the conclusion he remarked, "Mr. Tesla may accomplish great things, but he certainly will never do this. It would be equivalent to converting a steadily pulling force, like that of gravity into a rotary effort. It is a perpetual-motion scheme, an impossible idea." But instinct is something which transcends knowledge. We have, undoubtedly, certain finer fibers that enable us to perceive truths when logical deduction, or any other willful effort of the brain, is futile.[15]

To take up the challenge of building a spark-free motor, Tesla switched to the engineering course. However, electrical engineering was in its infancy and the course in Graz concentrated on civil engineering. Nevertheless, Tesla used his vivid imagination to pick away at the problem:

I started by first picturing in my mind a direct-current machine, running it and following the changing flow of the currents in the armature. Then I would imagine an alternator and investigate the progresses taking place in a similar manner. Next I would visualize systems comprising motors and generators and operate them in various ways. The images I saw were to me perfectly real and tangible.[16]

A diligent student, Tesla passed his exams at the end of the first year way ahead of his fellow students.

GOING OFF THE RAILS

In his second year at college, Tesla went off the rails. He was thrown out of school for gambling and, it was said, "womanizing." He disappeared from Graz without a word and moved to the Austrian province of Styria, but was deported back to Gospic. There he met and fell in love with a girl called Anna. She wanted to settle down and have a family; Tesla wanted to be an electrical engineer. Then his father died and he decided to straighten out.

I conquered my passion then and there and only regretted that it had not been a hundred times as strong. I not only vanquished but tore it from my heart so as not to leave even a trace of desire. Ever since that time I have been as indifferent to any form of gambling as to picking teeth.[17]

Tesla reported giving up excessive smoking and coffee drinking with similar ease. And he seems to have given up his passion for Anna too.

TELEPHONIC CONNECTIONS

With the support of two maternal uncles, Tesla went on to Prague University and signed up for courses in mathematics, experimental physics and philosophy. The intellectual ferment of Prague stimulated Tesla and, again, he put his mind to building a new type of electric motor, removing the commutator to eliminate sparking.

Eventually, the money from his uncles dried up. Tesla needed a job and he saw in the newspapers that one of Thomas Edison's agents, Tivadar Puskás, was setting up a telephone exchange in Budapest, having already built one in Paris. Puskás aimed to build telephone exchanges in major European cities. Originally, Alexander Graham Bell had only thought of installing his invention on private lines linking two locations, though taking Puskás' concept the Bell Telephone Company built the first experimental exchange in Boston, Massachusetts, in 1877. Tesla was taken on to work on the Budapest exchange[18] and recalled:

The knowledge and practical experience I gained in the course of this work, was most valuable and the employment gave me ample opportunities for the exercise of my inventive faculties. I made several improvements in the Central Station apparatus and perfected a telephone repeater or amplifier which was never patented or publicly described but would be creditable to me even today. In recognition of my efficient assistance the organizer of the undertaking, Mr. Puskás, upon disposing of his business in Budapest, offered me a position in Paris which I gladly accepted.[19]

Tesla and his friend Anthony Szigeti traveled to Paris where the Edison company was introducing Edison's incandescent lighting system.

TESLA'S EUREKA MOMENT

While he was in Budapest, Tesla suffered from depression. This was only relieved by taking walks with Szigeti. One day they were walking in the City Park reciting poetry when Tesla had what he later described as his Eureka moment.

The idea came like a flash of lightning and in an instant the truth was revealed. I drew with a stick on the sand the diagram shown six years later in my address before the American Institute of Electrical Engineers, and my companion understood them perfectly. The images I saw were wonderfully sharp and clear and had the solidity of metal and stone, so much so that I told him, "See my motor here; watch me reverse it." I cannot begin to describe my emotions.[20]

The idea Tesla had come up with was using a rotating electric field within the motor created by an alternating current. He did not patent his AC motor until 1903. But he did further experiments on it in 1883 and 1887, and the idea was still not fully worked out when he addressed the American Institute of Electrical Engineers in 1888. However, intellectually, Tesla had solved the problem Professor Pöschl had said was impossible.

TESLA THE TROUBLE-SHOOTER

In Paris, Tesla immediately found himself at odds with his colleagues. They were trained in the Edison way and made slow progress by trial and error. However, unlike the other engineers, Tesla had studied physics and mathematics and he could make calculations.

He continued to work on his AC motor, but he could not get any of Edison's men interested. The business making money at the time was delivering electric light rather than powering motors. The other problem was that, his design used six coils, rather than the three used in Edison's DC system. It would use much more copper. This was a major factor in the cost of new equipment at the time.

Tesla was also used as a trouble-shooter for the new Edison lighting stations in France and Germany. He was fluent in French and German, as well as Italian, English, Serbian, and Croatian.

He oversaw the illumination of an opera house in Paris, a theater in Bavaria and cafés in Berlin. After helping to develop an automatic regulator for Edison dynamos, he was sent to fix the illuminations of the central railway station in Strasbourg, which, since the Franco-Prussian War of 1870 – 71, had belonged to Germany.

There had been a problem there when the wiring had shorted, blowing out a section of the wall during a visit of Kaiser Wilhelm I and a German-speaking engineer was needed to sort it out.[21]

AC COMES ALIVE

In the station's powerhouse, there was a Siemens AC generator. This gave Tesla the opportunity to experiment with the prototype of one of his own designs of AC motors. "I undertook the construction of a simple motor in a mechanical shop opposite the railroad station, having brought with me from Paris some material for that purpose,"[22] he said.

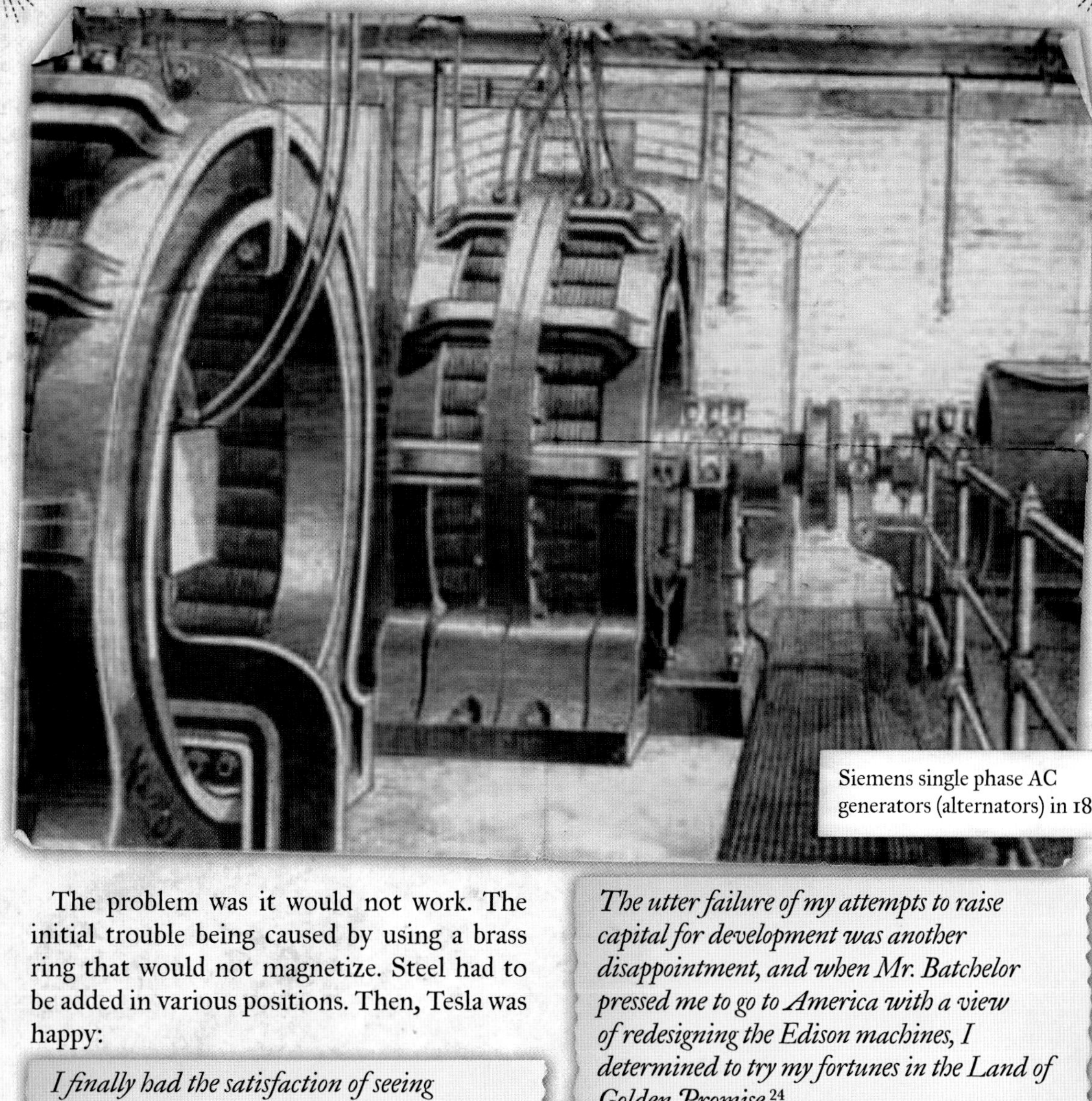

Siemens single phase AC generators (alternators) in 1885.

The problem was it would not work. The initial trouble being caused by using a brass ring that would not magnetize. Steel had to be added in various positions. Then, Tesla was happy:

I finally had the satisfaction of seeing rotation effected by alternating current of different phase, and without sliding contacts or commutator, as I had conceived a year before. It was an exquisite pleasure, but not to compare with the delirium of joy following the first revelation.[23]

Tesla tried unsuccessfully to raise money to back his invention in Strasbourg. He returned to Paris expecting a bonus for his work in Strasbourg. This did not materialize. He tried to find financial backing for his motor in Paris, again unsuccessfully. So he accepted Charles Batchelor's offer to go and work with the Edison company in New York. Tesla was frustrated:

The utter failure of my attempts to raise capital for development was another disappointment, and when Mr. Batchelor pressed me to go to America with a view of redesigning the Edison machines, I determined to try my fortunes in the Land of Golden Promise.[24]

TONGUES OF LIVING FLAME

Throughout his life, Tesla had a vivid imagination. But during his stay in Paris he suffered from intense hallucinations. He tried to describe them:

They were my strangest and most inexplicable experience. They usually occurred when I found myself in a dangerous or a distressing situation or when I was greatly exhilarated. In some instances I have seen the air around me filled with tongues of living flame. Their intensity, instead of diminishing, increased with time and seemingly attained a maximum when I was about twenty-five years old.[25]

TRAVELING TO NEW YORK

Tesla's journey to New York was not without incident. When he arrived at the Paris railroad station, he discovered his money and tickets were gone, not to mention his briefcase. With the train pulling out in great billows of steam, the frantic Tesla ran down the platform and scrambled on. The steamship company allowed him to board at the last minute only when no one else arrived to claim his berth.

I managed to embark for New York with the remnants of my belongings, some poems and articles I had written, and a package of calculations relating to solutions of an unsolvable integral and my flying machine. During the voyage I sat most of the time at the stern of the ship watching for an opportunity to save somebody from a watery grave. Later, when I had absorbed some of the practical American sense, I shivered at the recollection and marveled at my former folly.[26]

When Tesla sailed into New York harbor on board the *City of Richmond* on June 6, 1884, the first stones of the Statue of Liberty's pedestal were being hauled into place. He arrived with just four cents in his pocket and the address of a friend.

THE HIRED HAND

Walking north from the Battery, he passed a small machine shop where the foreman was trying to repair an electric motor. He had just given up the task as hopeless. Tesla stepped in and offered to help "without a thought of compensation."[27] He said: "It was a machine I had helped design, but I did not tell him that. I asked ... 'what would you give me if I fix it?' 'Twenty dollars' was the reply. I took off my coat and went to work, [and]... had it running perfectly in an hour."[28]

He moved on to 65 Fifth Avenue, Edison's headquarters. It was a large brownstone,

Edison Headquarters, 65 Fifth Avenue, New York.

fitted out with the latest lamps and electric chandeliers, aimed to entice wealthy customers. The top floor was given over to accommodation to unmarried employees, while another floor was given over to a night-school for new engineers. The company needed trained men. It was here that the elegantly attired Tesla had his "memorable meeting" with Edison, who smoked cigars, wore a shabby Prince Albert coat and generally looked like a Bowery bum.[29]

LIGHTING UP NEW YORK

In the summer of 1884, the Edison Electric Illuminating Company was busy lighting up the New York Stock Exchange, Brown Brothers & Company on Wall Street, the North British & Mercantile Insurance Company on William Street, the offices and pier of the New Haven Steamboat Company, and the *New York Commercial Advertiser*. Stand-alone generating plant, lighting a single building – usually a hotel or factory – were popular. By the fall of 1884, there were 378 in the US.[30]

But central stations, such as the one in Pearl Street with its limited range of half-a-mile in any direction, proved a harder sell. It was difficult to convince hundreds of businesses in the area that they needed electric power. By the end of 1884, there were only eighteen central stations in the country.[31]

THE AC SECRET

After fixing the lighting system on the SS *Oregon*, Tesla set about redesigning Edison's dynamos, replacing their long magnets with more efficient short cores, claiming that they gave three times the output for the same amount of iron. He kept quiet about his AC motor though, perhaps recalling the indifference of Edison's men in Paris. Once, however, he did attempt to bring up the subject with Edison himself.

All this time I was getting more and more anxious about the invention and was making up my mind to place it before Edison. I still remember an odd incident in this connection. One day in the latter part of 1884, Mr. Batchelor, the manager of the works, took me to Coney Island, where we met Edison in the company of his former wife. The moment that I was waiting for was propitious, and I was just about to speak, when a horrible-looking tramp took hold of Edison and drew him away, preventing me from carrying out my intentions ... That evening, when I came home, I had a fever and my resolve rose up again not to speak freely about it to other people.[32]

Otherwise Tesla and Edison got on well enough. Tesla told the tale of visiting Edison's office at 65 Fifth Avenue, when the great man was playing a game guessing weights.

"Edison felt me all over and said: 'Tesla weighs 152 pounds to the ounce,'" Tesla recalled, "and he guessed it exactly."[33]

He asked how Edison could guess his weight so accurately and was told: "He was employed for a long time in a Chicago slaughter house where he weighed thousands of hogs every day."[34]

VISION OF THE FUTURE

Tesla would occasionally dine with Edison, Batchelor and other of the company's top brass in a restaurant across the road from the showroom at 65 Fifth Avenue where they would swap stories and tell jokes. Afterwards they would go to a billiard room where Tesla would impress them with his bank shots and his vision of the future.

One night Tesla did get to discuss his ideas with Edison, pointing out that AC would free his central stations from their range limitation. His AC induction motors were superior to Edison's DC design, and would expand the market for electricity from his central station beyond lighting.

Edison, said Tesla, responded "very bluntly

that he was not interested in alternating current; there was no future to it and anyone who dabbled in that field was wasting his time; and besides, it was a deadly current whereas direct current was safe."[35]

THE POET OF SCIENCE

Companies that had grown up making arc lights were now moving into incandescent lighting, robbing Edison of valuable contracts. He struck back by going into arc lighting which was more suitable for street lighting or illuminating large spaces. Edison filed an arc lamp patent in June 1884 and left Tesla to work out the details. Tesla completed the job, but his system was shelved when Edison made a deal with a dedicated arc lighting company and, by then, larger incandescent bulbs suitable for lighting larger spaces had been developed.

The arc light companies, by and large, used high voltage alternating current. There had been several incidents where electrical workers had been shocked and killed, and Edison denounced AC as too dangerous for domestic use.[36] He insisted that the DC system he had built from the bottom up was the way ahead. For nearly a year Tesla worked for Edison:

During that period I designed twenty-four different types of standard machines with short cores and uniform pattern, which replaced the old ones. The manager had promised me fifty thousand dollars, on completion of the task, but when I demanded payment, he merely laughed. "You are still a Parisian," remarked Edison. "When you become a full-fledged American, you will appreciate an American joke."[37]

As it was, Tesla could not even get his salary of $18 a week increased to a modest $25. When another employee approached Batchelor about Tesla's raise, Batchelor replied: "The woods are full of men like him. I can get any number of them for $18 a week."[38]

This may have been a joke, but Tesla was outraged and quit. Edison was not sorry to see him go. He later dismissed Tesla as a "poet of science," saying: "His ideas are magnificent but utterly impractical."[39]

A New York blizzard highlights thousands of electric wires strung overhead in 1888.

CHAPTER 3

TESLA'S PYRO-MAGNETIC GENERATOR

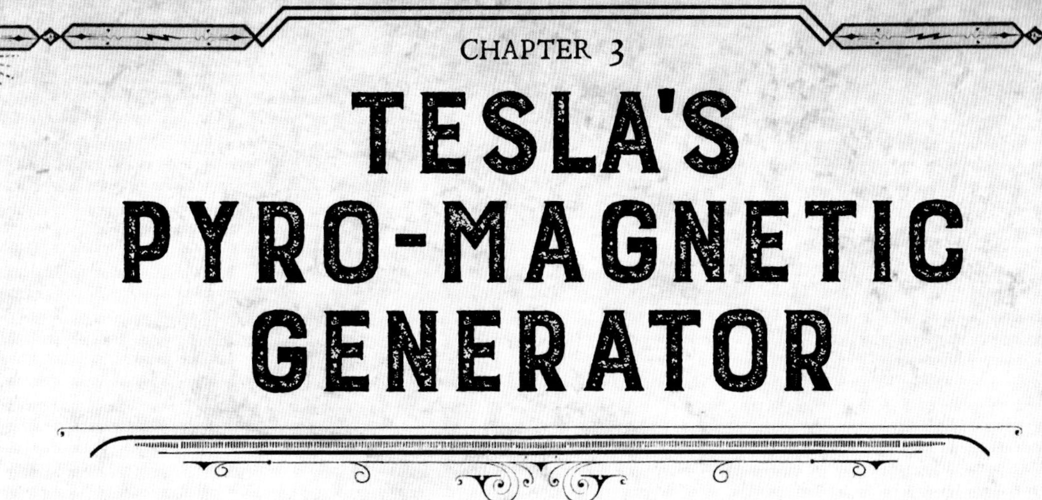

Despite the rift between the two men, Tesla was indebted to Edison. With Edison's former patent attorney, Lemuel Serrell, Tesla began patenting improvements to arc lights and dynamos. In Serrell's office, Tesla met B.A. Vail and Robert Lane. They set up the Tesla Electric Light & Manufacturing Company. Tesla remembered it like this:

> *Some people approached me with the proposal of forming an arc light company under my name, to which I agreed. Here, finally, was an opportunity to develop the motor, but when I broached the subject to my new associates they said, "No, we want the arc lamp. We don't care for this alternating current of yours."*[1]

The job in hand was lighting the streets and factories of Vail's hometown, Rahway, New Jersey. He set about installing powerful moonlight white arc lamps around the city. Meanwhile Tesla used the patents he had been granted to buy shares.[2] His first American patent was for a commutator on a dynamo electric machine on January 26, 1886, followed by an electric arc lamp on February 9, and a regulator for a dynamo electric machine on March 2.

When the electrification of Rahway was completed, *Electrical Review* featured it on the front page of its August 14, 1886 issue. Advertising in the same journal, the Tesla Electric Light & Manufacturing Company said it was "now prepared to furnish the most perfect automatic, self-regulating system of electric arc lighting yet produced." There was "no flickering or hissing of the lamps" under this "entirely new system of automatic regulation resulting in absolute safety and a great saving of power."[3]

Still Tesla believed that, once the lighting of Rahway was complete, his company could get on with the serious business of producing AC induction motors and huge electrical grids.

"The delay of my cherished plans was agonizing,"[4] he said.

The electrification was so successful that Vail and Lane decided to run the utility themseles, leaving no role for Tesla. He was bounced from the company leaving him penniless.

It was, he said, "the hardest blow I ever received. Through some local influences, I

Arc lighting in Madison Square, New York, 1882.

was forced out of the company, losing not only all my interest but also my reputation as engineer and inventor."[5] He said later: "I was free, but with no other possession than a beautifully engraved certificate of stock of hypothetical value."[6]

NO PLACE FOR ELECTRICAL SCIENCE

In the winter of 1886 – 87, America was beset with labor unrest and jobs were hard to come by. Too proud to return to Edison, Tesla found that, elsewhere, there was little call for his specialized electrical talents.

"My high education in various branches of science, mechanics and literature seemed to me like a mockery."[7] He tramped the streets looking for work:

> *There were many days when [I] did not know where my next meal was coming from. But I was never afraid to work, I went where some men were digging a ditch ... [and] said I wanted to work. The boss looked at my good clothes and white hands and laughed to the others ... but he said, "All right. Spit on your hands. Get in the ditch." And I worked harder than anybody. At the end of the day I had $2.*

EDISON AND LABOR UNREST

By 1886, Edison's business had grown so large that he no longer knew every employee personally. At his lamp factory out in New Jersey, eighty highly skilled filament sealers formed a union and "became very insolent," Edison said, "knowing that it was very impossible to manufacture lamps without them."[8] When they objected to the proposed firing of one of their members, Edison quickly designed thirty machines to automate their work, then fired the man as planned.

"The union went out," said Edison, later adding the punch line: "It has been out ever since."[9]

A committee at the Edison Machine Works on Goerck Street parlayed with their boss, Charles Batchelor, demanding the right to form a union and seek better pay and conditions. The Edison management, which paid average wages, agreed to reducing the workday from ten hours to nine, but they would have no truck with unions. The works were "to be run just as the managers decided and no interference whatever to be tolerated."[10] Nor would they accept the end of piecework. Batchelor explained to reporters, "If [a worker] loafs or gets drunk he loses his own time, not ours."

On May 19, the Edison workers went on strike. Soon afterward, Edison moved the machine works upstate to the quiet town of Schenectady, an easy ride on the New York Central Railroad. He wanted, he said, "to get away from the embarrassment of the strikes and the communists to a place where our men are settled in their own homes."

The Edison Machine Works on Goerck Street, New Jersey.

TESLA'S THERMO-MAGNETIC MOTOR

When a magnetized metal is heated it loses its magnetism at the "Curie temperature" – named after Pierre Curie, the co-discoverer of radium with his wife Marie. Iron loses its magnetism at 770° C (1,418° F), nickel at 360° C (680° F). Alloys such as nickel-iron may lose their magnetism at temperatures ranging from below room temperature to as high as 770° C, depending on the ratio of nickel to iron.

Tesla devised a small motor with one fixed magnet and a piece of metal attached to a flywheel and a pivot arm that pushed against a spring. At room temperature, the attraction between the magnet and the metal was enough to compress the spring and pull the metal into the flame of a Bunsen burner. This would heat up the metal. At the Curie temperature, it would lose its attraction. The spring would then push it back, turning the flywheel. With the metal now out of the flame, it would cool down to below the Curie temperature. It would then be attracted by the magnet again and the cycle would repeat.

A thermo-magnetic motor is capable of up to twenty strokes a minute because the Curie transformation point is sharp and the temperature of the magnetic properties of the metal disappear and recommence within a few degrees above or below the Curie temperature.

N. TESLA.

No. 396,121.

THERMO MAGNETIC MOTOR.

Patented Jan. 15, 1889.

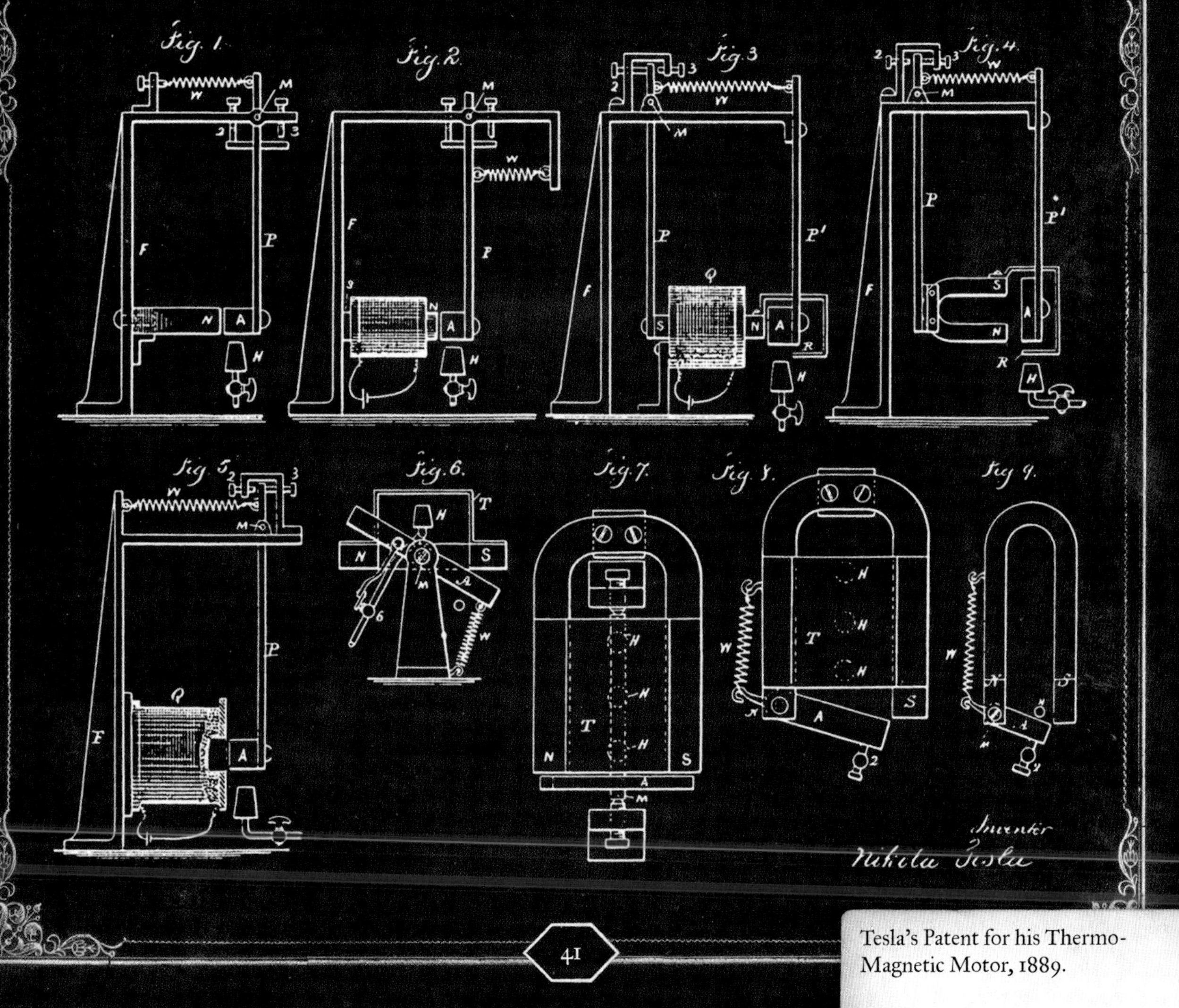

Tesla's Patent for his Thermo-Magnetic Motor, 1889.

SHEDDING BITTER TEARS

"I lived through a year of terrible heartaches and bitter tears, my sufferings being intensified by my material want,"[11] said Tesla. But his mind had not stopped working. Back in 1884, Edison had been experimenting with ways of producing electricity from burning coal or gas. It ended when an explosion blew the laboratory's windows out. However, Tesla figured out a simpler – and safer – way to do it, and in March 1886, he applied for a patent for his thermo-magnetic motor.

While Tesla was digging ditches, he told his foreman about his inventions. It seems that the foreman had been digging the ditches for the underground cables that connected Western Union's head office with the stock and commodity exchanges, and knew the engineer Alfred S. Brown (1836 – 1906) who was superintendent of Western Union's New York Metropolitan District.[12]

Brown would probably have known of Tesla from the article in *Electrical Review* and was impressed by his thermoelectric motor and the fervor with which he spoke about his AC motor.

FINDING INSPIRATION IN COLUMBUS

Eager to exploit Tesla's ideas, Brown contacted Charles F. Peck, a wealthy lawyer

THE EGG OF COLUMBUS

Tesla's demonstration was described in the pages of the *Electrical Experimenter*.

> *A rotating field magnet was fastened under the top board of a wooden table and Mr. Tesla provided a copper-plated egg and several brass balls and pivoted iron discs for convincing his prospective associates. He placed the egg on the table and, to their astonishment, it stood on end, but when they found it was rapidly spinning, their stupefaction was complete. The brass balls and pivoted iron discs in turn were sent spinning rapidly by the rotating field, to the amazement of the spectators. No sooner had they regained their composure than they delighted Tesla with the question, "Do you need any money?"*[13]

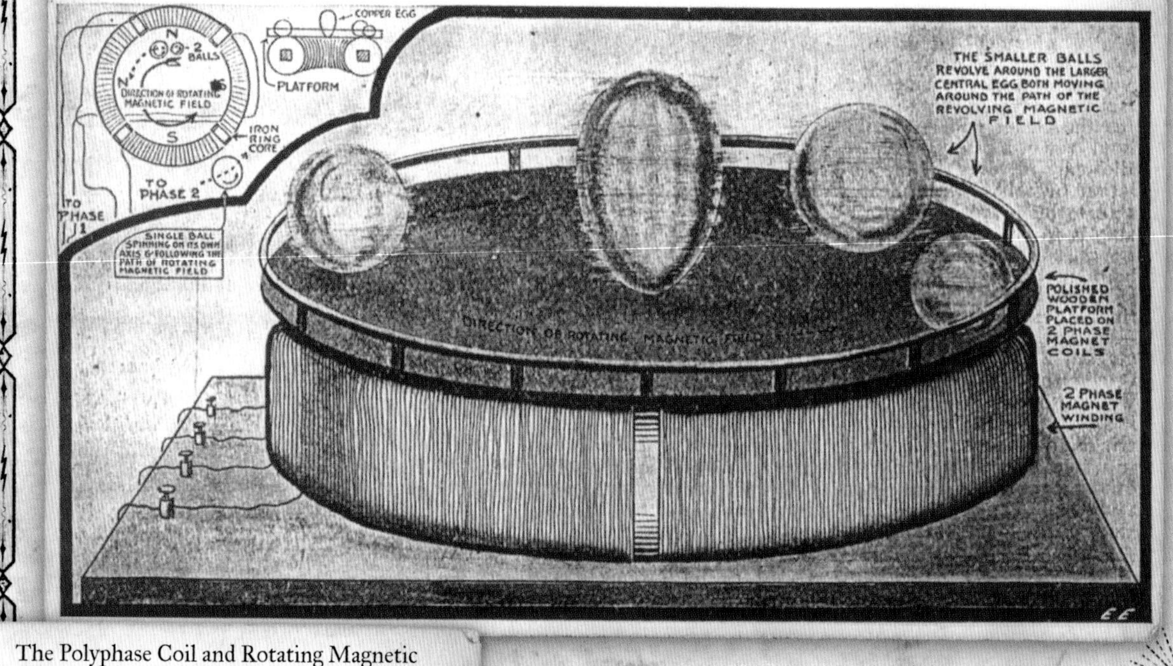

The Polyphase Coil and Rotating Magnetic Field caused the copper eggs to spin. Illustration from *Electrical Experimenter*, vol 22.

from New Jersey. However, Peck knew of the general prejudice against AC and refused even to witness some tests.

"I had an inspiration," Tesla said. Then he asked Peck: "Do you remember the Egg of Columbus?"[14] The story goes that Columbus was having dinner with some Spanish nobles who mocked him. So he challenged them to stand an egg on its end. They all tried and failed.

Columbus then took the egg, tapped it lightly on one end, slightly cracking the shell and denting it, so it would stand upright. As a result he was granted an audience with Queen Isabella and won her support for his voyage to what turned out to be the New World.

"And you plan to balance an egg on its end?" asked Peck.

"Yes," said Tesla, "but without cracking the shell. If I should do this, would you admit that I had gone one better than Columbus?"

Intrigued, Peck said he would. "And would you be willing to go out of your way as much as Isabella?" asked Tesla.

"I have no crown jewels to pawn," Peck conceded. "But there are a few ducats in my buckskins, and I might be able to help you to an extent."[15]

After the meeting, Tesla took a hard-boiled egg to a blacksmith and got him to cast one in copper. He then placed four coils under the top of a table to create a magnetic rotating field. The demonstration took place the next day.

When Tesla turned on the current, the magnetic field induced an electric current in the copper shell of the egg, which produced its own magnetic field. As this reacted against the magnetic field from the coils, the egg began to spin. As it spun faster and faster, the egg stopped wobbling and stood on its end. Peck was astounded. Not only had Tesla trounced Columbus, he had demonstrated the principle of his AC motor, impressing a man who was skeptical about the use of alternating current.

STILL SHARING PATENTS

Together Tesla, Brown and Peck formed the Tesla Electric Company.[16] Tesla agreed that the patents would be split fifty-fifty. The first was filed on April 30, 1887. In fact, the money generated from the six patents Tesla filed on his AC motor and power transmission was split three ways. Tesla got a third. Brown and Peck split another third, and a third was reinvested. Tesla also received a salary of $250 a month, leaving him free to follow wherever his imagination led him. The money was raised by Peck and he would cover the cost of the patents. Brown's role was to provide his technical expertise. Tesla sent for his old friend, Anthony Szigeti, to assist him. Szigeti arrived in New York the following month.

Brown found them a laboratory at 89 Liberty Street above a printing shop.[17] During the day the printer used a steam engine to power the presses. At night it provided electric power for Tesla's experiments.

First, Tesla developed his thermo-magnetic motor into the pyro-magnetic generator, using the same principles. Tesla believed this was a great invention and worked on it from the fall of 1886 to the late summer of 1887, but it did not work very well as, to generate significant amounts of electricity the temperature would have to rise and fall rapidly. As it was, the core retained its latent heat. Tesla applied for a patent, but the application was denied.

Tesla began to fear that Peck and Brown might abandon him just as Vail and Lane had done. But Peck still had every confidence in his inventor. Tesla recalled:

I met Mr. Peck just at the door of the building in which he had his office, and he spoke to me in a very kind way and said, "Now do not be discouraged that this great invention of yours is not panning out right; you may bring it in a success after all. Perhaps it would be good if you would switch to some other of your ideas and drop this for a while. I have had an experience that this is a very good plan." I came back encouraged.[18]

TESLA'S PYRO-MAGNETIC GENERATOR

In his pyro-magnetic generator, Tesla combined the principles he had employed in his thermo-magnetic motor and Faraday's laws of electromagnetic induction. So if a magnet was alternately heated up and cooled down, it would induce a current in a nearby conductor.

Tesla took a large horseshoe magnet. Across its poles he laid a thermally insulated metal box, containing a number of hollow iron tubes which had coils of wire wrapped around them. Because they were sitting on the horseshoe magnet, these iron tubes were magnetized.

Under the center of the tubes was a firebox which heated them. Above there was a boiler which produced steam that circulated through the iron pipes via a valve. The coal fire in the firebox heated the iron tubes until they reached about 600°C (1,112°F) and glowed a dull red. At this temperature, the tubes would become demagnetized and the collapsing magnetic field would induce an electric current in the coils. Then the valve was opened, and steam at 100°C (212°F) would circulate through the tubes, cooling them. As the tubes became magnetized again, electric current would again be induced in the coils. The heating and cooling induced currents moving in different directions in the coils, so the pyro-magnetic generator produced AC.

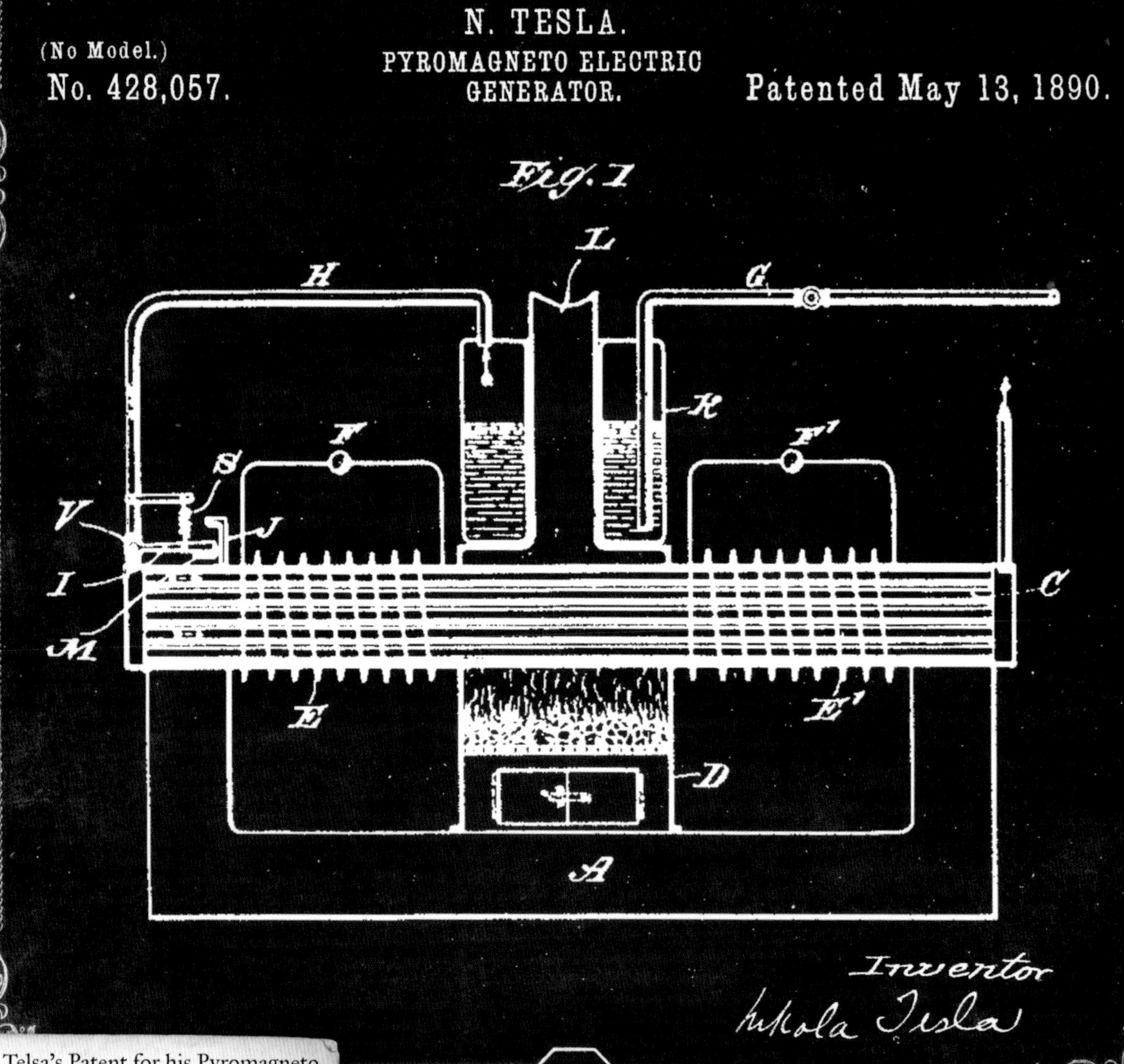

Telsa's Patent for his Pyromagneto Electric Generator, 1890.

IMAGINING THE AC MOTOR

Despite this setback, Peck encouraged Tesla to continue inventing and his mind turned back to the AC motor he built in Strasbourg.

Tesla began a vigorous schedule that frightened those around him. On many occasions, he drove himself until he collapsed, working around the clock, with few breaks. Instead of a single coil, he wound four coils around a laminated ring. Two separate AC currents were fed to the pairs of coils on opposite sides. If these two currents were 90° out of phase – that is, when one was at its maximum positive value, the other was at its maximum negative value – a rotating electrical field was produced. To Tesla's delight, the rotor – initially a shoe polish tin balanced on a pin – began to spin.[19]

From this prototype, Tesla and Szigeti produced two full-scale motors. They were, Tesla said, "exactly as I had imagined them. I made no attempt to improve the design, but merely reproduce the pictures as they appeared to my vision and the operation was as I expected."[20]

According to Tesla's longstanding friend and first biographer John O'Neill:

> *Tesla produced as rapidly as the machines could be constructed three complete systems of AC machinery – for single-phase, two-phase, and three-phase currents – and made experiments with four- and six-phase currents. In each of the three principal systems he produced the dynamos for generating the currents, the motors for producing power from them, and transformers for raising and reducing the voltages as well as a variety of devices for automatically controlling the machinery. He not only produced the three systems but provided methods by which they could be interconnected and modifications providing a variety of means of using each of the systems.*[21]

He also calculated, in fundamental fashion, the mathematics behind these inventions.

TESLA THE TRAILBLAZER

The editor of *Electrical World*, Thomas Commerford Martin, encouraged Tesla to write his first article about his invention. Martin, who had worked for Edison in the 1870s, became a prominent advocate of Tesla. Martin described the 35-year-old Tesla:

> *He is tall and spare with a clean-cut, thin, refined face, and eyes that recall all the stories one has read of keenness of vision and phenomenal ability to see through things. He is an omnivorous reader, who never forgets; and he possesses the peculiar facility in languages that enables the least educated native of eastern Europe to talk and write in at least half a dozen tongues. A more congenial companion cannot be desired for the hours when one "pours out heart affluence in discursive talk," and when the conversation, dealing at first with things near at hand and next to us, reaches out and rises to the greater questions of life, duty and destiny.*[22]

Tesla applied for patents on his new AC motors. They were granted on May 1, 1888. Martin then arranged for Professor William Anthony (1835 – 1908) of Cornell University to test his motors for efficiency. Tesla also visited Cornell to demonstrate his motors to the academics there.

Martin and Anthony then persuaded Tesla to present a paper to the newly formed American Institute of Electrical Engineers. Written hastily the night before in pencil, Tesla delivered his ground-breaking paper "A New System of Alternate Current Motors and Transformers" to the AIEE on May 15, 1888. Even then, on the advice of his patent lawyers, he held back many of the details. Already, Tesla, Brown, and Peck were in negotiations with potential buyers, including George Westinghouse of Pittsburgh.[23]

D
E
B
C
J
G
N
L
F
K
M
D
E
K
J
L
P
P
H
N
O
O

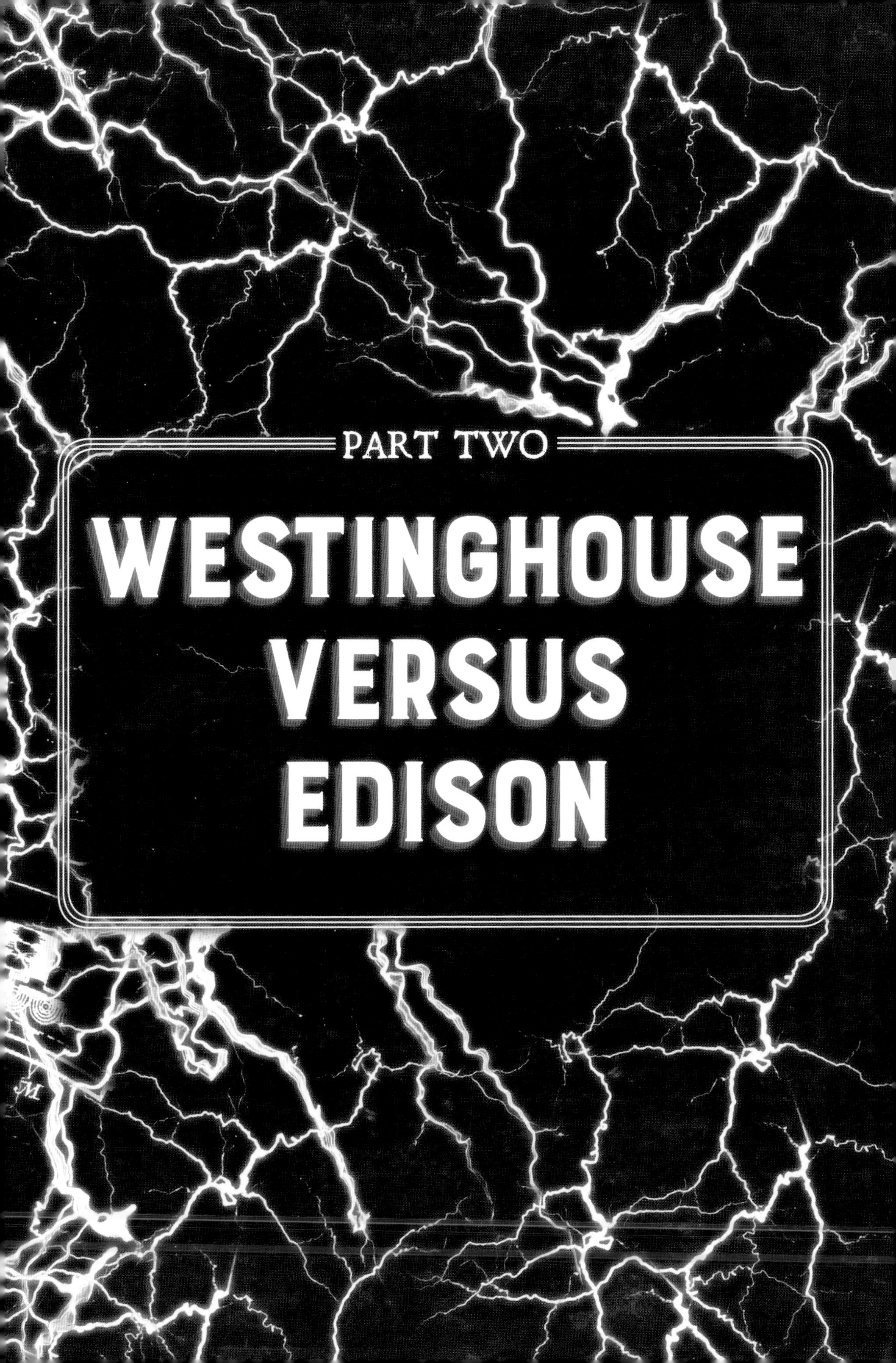
PART TWO
WESTINGHOUSE
VERSUS
EDISON

CHAPTER 4

WESTINGHOUSE AND AC

Along with Edison and Tesla, there was a third key player in the battle between AC and DC. This was George Westinghouse (1846 – 1914). He was a descendant of the aristocratic Russian von Wistinghousen family. His father was also an inventor with six patents for farming machinery to his name.

In his father's machine shop in Schenectady, New York, the young Westinghouse experimented with batteries and Leyden jars – glass jars coated with metal foil, used for storing electrical charge. At fifteen, he made his first invention, a rotary steam engine. After serving in both the US Army and Navy during the American Civil War, Westinghouse attended the nearby Union College, but soon dropped out. He said:

My early greatest capital was the experience and skill acquired from the opportunity given me, when I was young, to work with all kinds of machinery, coupled later with the lessons in that discipline to which a soldier is required to submit.[1]

In 1865, he patented his rotary engine and the "car replacer" – a device for putting derailed freight cars back on the tracks. Soon after, he designed a reversible cast-steel frog which prolonged the life of railroad track switches. However, the railroads quickly improved on this and filed their own patents.

Having been involved in a near collision on the railway, he put his mind to improving the braking system which, until then, had depended on a brakeman on every carriage. His first system, using steam, proved impractical. But then, in 1869, he came up

George Westinghouse (1846 – 1914).

THE WESTINGHOUSE AIR BRAKE

The air-operated brake was invented by George Westinghouse in 1869 and became standard equipment on railroad cars in 1872. Before that, the brakeman had to apply mechanical brakes manually in response to a whistle from the engineer. Train wrecks were commonplace. With the Westinghouse automatic air brake, the engineer in the cab of the locomotive was in control of stopping the train.

Typically an air compressor was located on the locomotive. The air went to a brake valve controlled by the engineer, then via pipes on the cars and rubber connectors to a control valve on each car.

When pressure was applied to the control valve, a piston was depressed, diverting air into a reservoir, blocking any flow into the brake cylinder. If the engineer closed the brake valve, the pressure in the line was reduced. A spring forced the piston in the control valve in each car upwards. This allowed air to flow from the reservoir into the brake cylinder, pushing the brake pad against the wheel, applying friction to slow the wheel, and stop the train.

The advantage of the Westinghouse system was that, if the airline connecting the cars was broken – when the train separated for example – the pressure was lost and the brakes were automatically applied, stopping a runaway train.

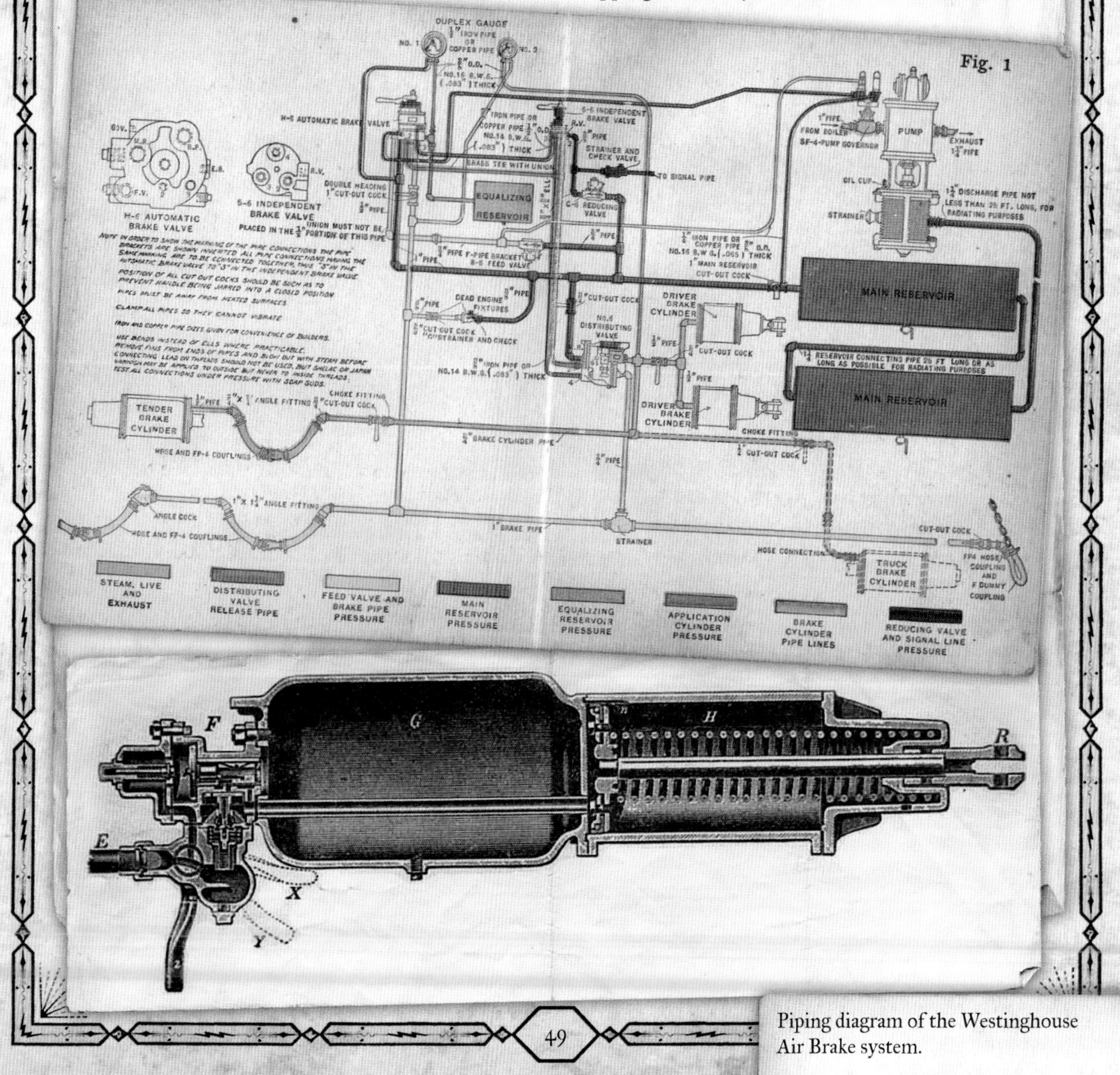

Piping diagram of the Westinghouse Air Brake system.

with air brakes that soon became standard in the US and Europe.

Having suffered at the hands of the railroads previously, he refused to grant the railroads licenses and monopolized the manufacture of the air brakes from his works in Pittsburgh. He threatened to sue anyone who tried to infringe his patents.

AUTOMATING ELECTRIC SIGNALS

Setting out to sell the products of the Westinghouse Air Brake Company in England, Westinghouse began buying up patents that used electric circuits to activate signals on the railroads. In the US, oil lamps were used in signaling systems, so Westinghouse established the Union Switch and Signal Company and hired the inventor William Stanley Jr.

After one semester at Yale, Stanley had quit to go to New York where he worked with the Swan Incandescent Electric Light Company and Hiram Maxim, the inventor of the machine-gun. Stanley had invented an incandescent lamp using a carbonized silk-filament and a self-regulating DC dynamo. Westinghouse bought the patents from Swan for $50,000 and paid Stanley a salary of $5,000 a year. Stanley would also receive ten percent of the profit of any of his patented inventions the company manufactured and sold.[2]

They concentrated on setting up a commercial light bulb facility in Pittsburgh and developing a DC system for Westinghouse. The new Westinghouse system had its debut at the 1884 Philadelphia Electric Exhibit. *Electrical World* that September reported: "The company are now prepared to do business. Their display comprises electric motors as well. An ingenious arrangement of the lamps is shown by which when one goes out another is switched into circuit, and by which also a bell announcing the occurrence can be rung at any chosen place."

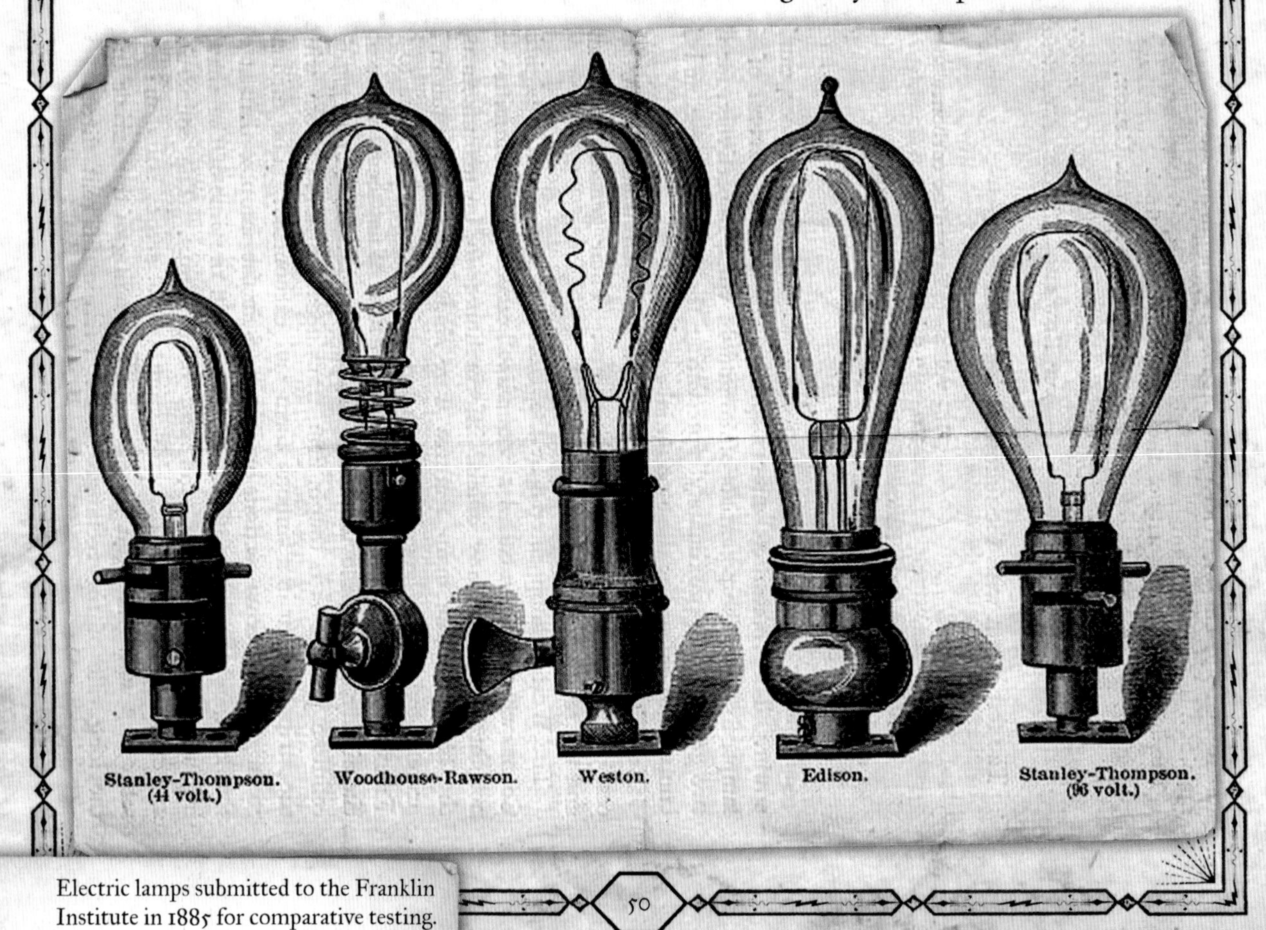

Electric lamps submitted to the Franklin Institute in 1885 for comparative testing.

SHUNNING THE LIMELIGHT

Westinghouse was a very different character from Edison. He was happy to buy in patents, while Edison would only exploit patents that he had filed himself. Edison was also the darling of the press, always launching his new products with a huge song and dance. In contrast, Westinghouse usually refused requests for interviews and stories.

"If my face becomes too familiar to the public, every bore or crazy schemer will insist on buttonholing me,"[3] he said. Even when he did grant an interview, Westinghouse rarely made for good copy.

However, in private he was said to be charismatic. One biographer wrote: "With his soft voice, his kind eyes, and his gentle smile, he could charm a bird out of a tree. It is related that in a knotty negotiation it was suggested to the late Jacob H. Schiff, then the head of a great banking house, that he should meet Westinghouse. 'No,' said the astute old Jew, 'I do not wish to see Mr. Westinghouse; he would persuade me.'"[4]

Behind the scenes, Westinghouse could also be blunt to the point of being offensive. But his public persona was reserved and serious. He had a solid reputation for making novel technologies work in the real world. His major inventions – railroad air brakes and automated signaling systems – had improved the safety and productivity of the nation's most important industry. Nevertheless, they lacked the glamor and glitz of Edison's discoveries, the talking phonograph and the incandescent light. When visiting the US, the French actress Sarah Bernhardt had begged to meet "*le grand* Edison," never "*le grand* Westinghouse."

WESTINGHOUSE GOES AC

Initially, Westinghouse thought of developing a DC system. However the market was already overcrowded and DC had its limitations. To give coverage to even a relatively small city, numerous small power stations would have to be situated near to the homes and factories they served.

In 1885, Westinghouse read an article in the British magazine *Engineering* that described an AC system on display at the Inventions Exhibition in London. It incorporated what was then known as a secondary generator – soon dubbed a transformer – that stepped down the high voltages used to power arc lamps to the lower voltages needed to feed individual incandescent light bulbs. This meant that high voltages could be used on transmission lines, then reduced to a safer low voltage before it entered a house, shop, office, or factory.[5]

Providing AC power over a wide area would bring major economies of scale. But there were still considerable disadvantages that had to be overcome. No one had yet developed a serviceable AC motor to provide motive power for streetcars and factories, nor had a meter been devised to measure how much each individual consumer used. This was one of the reasons Edison had dismissed AC as "not worth the attention of practical men."[6] As a businessman, if he provided electricity he wanted to be able to charge for it.

IMPORTING THE TECHNOLOGY

Westinghouse's employee, Guido Pantaleoni, was in Italy with instructions to track down the inventors of the "secondary generator," Lucien Gaulard and John Gibbs.

It happened that Gaulard and Gibbs were in Turin at the time. They had been exhibiting their system for some time and were currently installing a fifty-mile transmission circuit powered by a dynamo driven by a waterfall in the Alps, lighting the railroad stations in Lanzo, Veneria Reale, and Turin.

Pantaleoni had his doubts about this venture and consulted the German inventor and industrialist Werner von Siemens. He assured Pantaleoni that the use of AC was "pure humbug."[7] However, he also consulted the Ganz Company in Budapest where it is

PITTSBURGH IN 1885

Westinghouse built his works in Pittsburgh because the abundance of coal there made the casting of his frogs cheaper. In November 1885, Reginald Belfield, an employee of Gaulard-Gibbs, arrived in Pittsburgh with one of the first transformers. He would have witnessed a place engulfed by the fires of industry. An observer at the time described the city:

Pittsburg [as it was spelled from 1890 to 1911] of today looks in the distance like a huge volcano, continually belching forth smoke and flames. By day a great pall rests over the city, obscuring the sun, and by night the glow and flash of the almost numberless iron mills which fill the valley and cover the hillsides, light the sky with a fiery glare. This great workshop of the modern Cyclops is one of the most important manufacturing centers of the country, and embodies our most valuable interests in iron and steel manufactures ... Though the suburbs of the city are beautiful and contain many charming residences, the aspect of the city itself is grimy and gloomy, in spite of the noble business blocks and open, spacious streets.[8]

English novelist Anthony Trollope called it "the blackest place ... I ever saw." Biographer James Parton said, "every object is black. Smoke, smoke ... everywhere smoke ... Hell with the lid taken off." And visiting English anthropologist Herbert Spence said, "Six months residence here would justify suicide."[9]

Not only was there coal and smoke, there was gas and flames too. Seeking to exploit local reserves of natural gas, Westinghouse almost burnt down his own house, "Solitude," in the suburb of Homewood.

Smokestacks polluting Pittsburgh, *c.* 1885.

thought that Tesla found inspiration when he lived there. At the time, Ganz were developing the Gaulard-Gibbs' system – to the point of infringing their patents.

In 1885, Westinghouse imported a transformer made by Lucien Gaulard and John Gibbs in London, which William Stanley would experiment with. At the same time, Westinghouse bought the US patent rights of a Siemens AC generator designed to run arc lighting.

REBUILDING THE FUTURE

The arrival of the transformer itself was problematic. Preferring to work on his own inventions, William Stanley had moved out to a laboratory in Great Barrington in rural Massachusetts, while Reginald Belfield was left to unpack the equipment from its wooden crate. Belfield recorded the event:

> *The Gaulard and Gibbs apparatus from England was sent in a very unsatisfactory condition. The work in the various parts was very faulty; portions not properly soldered only held together by the flux. The impression this gave was so bad that Mr. Pantaleoni was going to telegraph to London canceling the arrangement that had been made. Here Mr. Westinghouse stepped in and with his well-known sympathy towards a new invention, gave me another chance, for which I was most grateful, so that I had to practically reconstruct the entire apparatus and build most of it anew. This made a great deal of hard work but the result was satisfactory, inasmuch as Mr. Westinghouse determined to take up the apparatus. During this period I was staying at "Solitude," and had occasion every night to discuss the matter with him, and the future development of the system.*[10]

The Gaulard-Gibbs transformer, according to Belfield, "consisted of a bundle of iron wire forming the magnetic circuit surrounded by a large number of copper discs, having a hole in their centers, one to each turn of both primary and secondary, and each soldered to its neighbor, this multitude of soldered joints, each a source of trouble, was most impractical and very expensive to make."[11]

Westinghouse took the time to disassemble the transformer and discover how it worked. Belfield went on:

> *Those who knew Mr. Westinghouse will fully realize the great energy he threw into this question ... In an astonishingly short time the absolutely uncommercial secondary generator was converted into the modern transformer.*[12]

The Gaulard-Gibbs transformer.

WESTINGHOUSE ELECTRIC COMPANY

The Westinghouse Organization were still involved in DC generation and few in the company, apart from Westinghouse himself, were converted. William Stanley did however, recognize the merits of AC and Westinghouse renewed his contract. Stanley's laboratories in Great Barrington were adjusted to handle AC work and Reginald Belfield was sent there, to help make the new Westinghouse H-shaped transformers.

In 1886, Westinghouse set up the Westinghouse Electric Company. But by early March, Edison had moved into Great Barrington, lighting an establishment there using DC. The Siemen's AC generator powered by a steam engine in Stanley's lab on Main Street fired into action soon after, lighting a store owned by Stanley's cousin. It was noted that, under electric light rather than gaslight, you could distinguish between blue and green. However, Stanley took care not to mention to the press that the system he was using was any different from Edison's. Indeed, the transformers were kept out of sight, under lock and key.

Stanley then developed his own generator for a demonstration plant at the Union Switch and Signal Company in Pittsburgh with an electrical line strung out three miles to East Liberty. That fall, Westinghouse had his commercial AC incandescent system on the market.

THE WESTINGHOUSE ELECTRIC COMPANY

OF

PITTSBURGH,

Manufacturers of Isolated Incandescent Plants, and Contractors for Central Stations.

It is believed that the advantages of our System place us beyond competition.

Capital investing for dividends will do well to close no contracts until our proposals are considered.

Address
THE WESTINGHOUSE ELECTRIC CO.,
Pittsburgh, Pa.,
or
WESTINGHOUSE, CHURCH, KERR, & CO.,
17 Cortlandt Street,
New York.

16 Candle-power Lamp.

Newspaper advertisement for Westinghouse Electric Company, 1886.

ELECTRICITY FOR SALE

Westinghouse's first customer was the four-story department store, Adam, Meldrum & Anderson, on Main Street, Buffalo, New York. On the front page of the *Buffalo Commercial Advertiser* of November 27, 1886, the store announced that its latest range of goods were to be lit by 498 Stanley lights run by the Westinghouse system.

"There is no odor, no heat, no matches, no danger," its ad said. "We were the first business house in the city to adopt the plan of lighting our stores by incandescence ... The appearance is brilliant in the extreme. The light is steady and colorless. Shades can be perfectly matched. Come and see the grandest invention of the nineteenth century."[1]

Two days later, the store was packed – not with shoppers but a crowd that had come to

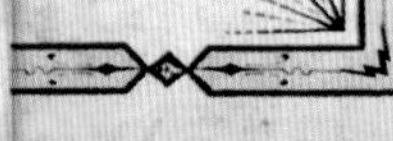

witness one of the wonders of the nineteenth century. A few of them had come from other lighting companies, though Westinghouse had made no announcement that he was using AC. Nevertheless, twenty-seven other customers signed up.

Fearing his new rival in the electrical field, Edison wrote: "Just as certain as death, Westinghouse will kill a customer within six months after he puts in a system of any size. He has got a new thing, and it will require a great deal of experimenting to get it working practically. It will never be free from danger ... I cannot for the life of me see how alternating current high-pressure mains – which in large cities can never stop – could be repaired. Whereas the mains of the direct current would not produce death if a man received a shock."[2]

DEATH BY WIRE

In March 1888, a blizzard hit New York, bringing down many of the overhead wires. The following month, while these were being restrung, a young boy grabbed hold of one of the dangling wires and was electrocuted. The cable belonged to the U.S. Illuminating Company that supplied high-voltage electricity for arc lighting.[3] Suddenly "death by wire" became the hot topic in the newspapers.

The following month, a lineman for the Brush Electric Company, another supplier of high-voltage for arc lighting,[4] was also electrocuted – though that was not the term used then.

Despite the growing fear of high voltages, Westinghouse signed up to build sixty-eight AC stations, while another of Edison's rivals, Thomson-Houston, were building another twenty-two, using Westinghouse transformers. By then, Edison's DC was underway with 121 power stations from Grand Rapids, Michigan, to Birmingham, Alabama, and Edison was still confident that he would see off the competition from AC. However, Edward Johnson, one of Edison's senior lieutenants, was concerned that, without AC, they would "do no small-town business, or even [make] much headway in cities of minor size."[5] Other Edison employees were examining the Ganz system and urged Edison to buy into it. But Edison had genuine fears that further deaths from poorly designed high-voltage AC systems would impede the adoption of electric power.

Death by wire of an electric company lineman, watched by crowds of onlookers.

COPPER PRICE ROCKETS

Another consideration was the escalating price of copper when, spotting the growing demand by the electrical industry, a syndicate tried to corner the market. This favored AC as less copper was required in its high-voltage currents, as the current they carried was correspondingly lower.

The Edison Electric Light Company was reeling under the costs, when Edison was approached by a member of the New York State Death Commission, who were looking into an alternative to hanging as a punishment for capital crimes. Although Edison opposed capital punishment, he wrote back saying that the quickest, most painless death could be accomplished by the use of electricity:

> *The most suitable apparatus for the purpose is that class of dynamo-electric machine which employs intermittent currents. The most effective of those are known as 'alternating machines,' manufactured principally in this country by Geo. Westinghouse ... The passage of the current from these machines through the human body even by the slightest contacts, produces instantaneous death.*[6]

THE BATTLE OF THE CURRENTS BEGINS

In February 1888, Edison published *WARNING!*, an eighty-four-page attack on Westinghouse and AC. "It is a matter of fact that any system employing high pressures, i.e. 500 to 2,000 units, jeopardizes life."[7]

The pamphlet then detailed numerous gruesome deaths and injuries caused by high voltages, singling out for particular admonition the "Westinghouse alternative current." On the other hand, Edison insisted: "There is no danger to life, health, or person, in the current generated by any of the Edison dynamos ... the wires at any part of the system, and even the poles of the generator itself, may be grasped by the naked hand without the slightest effect."

A WARNING FROM THE EDISON ELECTRIC LIGHT CO.

The cunningly-phrased announcement broadcast over the wires of the Associated Press has failed to immortalize the name of George H. Westinghouse, as the inventor of the vaunted system of distribution which is today recognized by every thoroughly-read electrician as only an ignis fatuus, in following which the Pittsburg company have at every step sunk deeper in the quagmire of disappointment. The enticing glimmer of this transitory will-o'-the-wisp led them to negotiate, through a so-called "expert" agency, for an assignment of a claim then in the Patent Office. The curious coincidence of the couching of the news of their acquisition in mysterious and ambiguous language, in combination with the peculiar and unusual scarcity of the specifications of this patent, at the time of its issue, have been naturally enough widely commented upon. The fact that the electrician of the company gave some months of experiment to the system in an isolated locality, without even as much as being able to tell how many lamps he got to a horse-power expended; the fact that while the company was using an exhibition in Pittsburg as a platform by which to mount to public favor, their decrepit two-wire installation in East Liberty (Pittsburg) was transmogrified into a burlesque upon Mr. Edison's three-wire system; the fact that the Gaulard and Gibbs system, covered by this assignment, has never in any instance been worked to a successful issue by the Westinghouse Company, and cannot be so worked by anybody; the fact that it is not even a fair representative of the electrical principles sought to be embodied: all these show, as if by a burst of sunlight, the ridiculous and comfortless situation of those who, lured by the fascinating gleam of this phosphorescent decoy, have by a great effort reached, not the electrical palace which their fancy has pictured, but a whited sepulcher, full of dead men's bones.[8]

The pamphlet also sought to show that the Westinghouse system was not financially viable and the recent hike in the company's stock was a bubble. At the same time, Edison went on the attack in the courts with Westinghouse and others over infringement of his 1879 light bulb patent. Westinghouse filed a countersuit as he had bought the patents of men who been working on the incandescent light bulb long before Edison.

TESLA'S AC FIXATION

Meanwhile, power stations had sprung up across America and Europe to provide electric light at night. In the 1880s, the owners saw electric motors as a way to sell power to factories and streetcar lines during the day. However, most of the power stations produced DC and Brown and Peck were becoming a little dubious about Tesla's fixation with AC.

Other inventors had used AC to power arc lights. This was particularly popular in Europe where experimenters found they could raise or lower the voltage of an alternating current using primitive transformers. Engineers at the Ganz Company found that, at a high voltage, electricity could be distributed over long distances using thin copper wires. Then, to make it safe to use in the home, it would be stepped down using a transformer.

HOW TRANSFORMERS WORK

The first transformer was built by Michael Faraday in 1831. Two coils of insulated wire were wound around a single iron core. When an electric current was passed through the "primary," it created a magnetic field. This induced an electric current in the "secondary" coil. The voltage was stepped up or stepped down according to the ratio of the number of turns of wire in primary and secondary. However, induction only works when the electrical current is being switched on and off again, so an alternating current rather than a direct current must be used. William Stanley said:

> *I have a very personal affection for the transformer. It is such a complete and simple solution for a difficult problem. It so puts to shame all mechanical attempts at regulation. It handles with such ease, certainty, and economy vast loads of energy that are instantly given to or taken from it. It is so reliable, strong, and certain. In this mingled steel and copper, extraordinary forces are so nicely balanced as to be almost unsuspected. This equilibrium is remarkable ... It has to be remembered at this time, that is, in 1885, there were no alternating current machines built in America. The only transformers or induction coils that I knew of were three or four or more of the Gaulard type that had been imported from England.*[9]

The transformer is a vital component of any power distribution system as transmission losses are much smaller when the voltage is higher – as less current is needed to convey the same amount of energy. So electricity generated at a power station is stepped up in voltage using a transformer before it reaches the transmission lines. Then, at its destination, it is stepped down for use in the home or factory.

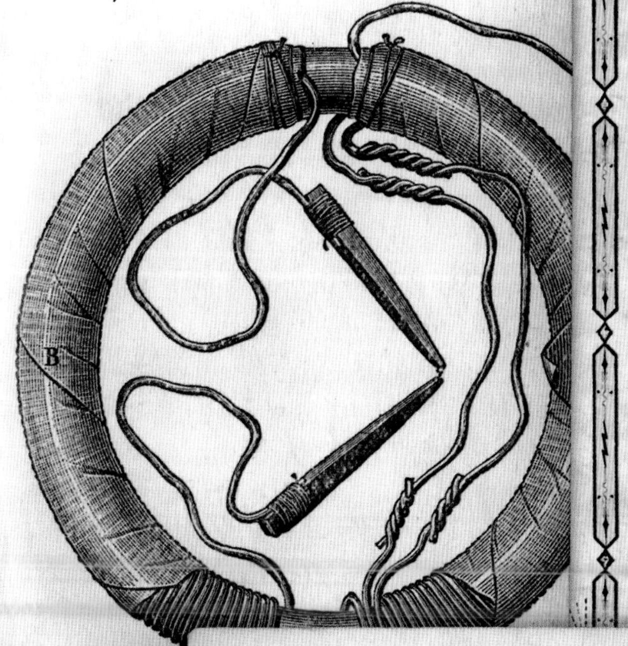

Drawing of the first transformer, built by British scientist Michael Faraday in 1831.

MICHAEL FARADAY
(1791 – 1867)

The son of a blacksmith, Faraday was born in Newington, Surrey in England, now part of south London. He received only a rudimentary education, largely at Sunday school. However, at fourteen he was apprenticed to a bookbinder and seized the opportunity to read the books brought in.

After reading an article on electricity in the *Encyclopaedia Britannica*, he began experimenting. He attended a lecture by Sir Humphry Davy, who was impressed by his note-taking and, later, offered him a job as his laboratory assistant. Davy was revolutionizing chemistry at the time.

Faraday worked as a chemist, established the scientific basis of metallurgy, and improved the quality of optical glass for telescopes. At the Royal Institution, he began a series of experiments into the nature of magnetism and electricity, and in 1831 discovered electromagnetic induction. He then built the first dynamo.

He continued research into electrochemistry and in 1839 developed a general theory of electrical action. Six years later he speculated on the electromagnetic nature of light. He developed new ideas about the nature of space and energy, marking the beginnings of field theory which was then developed mathematically by James Clerk Maxwell.

CHAPTER 6

TESLA DEMONSTRATES AC

Meanwhile, as Edison and Westinghouse were publicly slugging it out, back at the Tesla Electric Company, Anthony Szigeti had quietly arrived in New York in May 1887 to join Tesla in his Liberty Street lab. With Brown and Peck watching over them, the two old friends began work on the type of AC induction motors that Tesla had drawn in the sand to Szigeti in Budapest's City Park. Although in the intervening years, Tesla had not committed a single line of the design to paper, they worked exactly as he envisaged.[1]

They built several variants on the basic design, using a single-phase alternating current, two-phase and three-phase AC, both copper and iron. They also built a dynamo, again without a commutator and a transformer to step the power up and down. Within six months, they had a complete polyphase AC system. Tesla submitted a patent application on October 12, 1887.

The details of the whole system proved too much for the Patent Office to digest. They sent it back with a request that it be broken down and over the next two months, Tesla began filing what would eventually become forty patents covering his revolutionary induction motor and the AC system.[2]

KEEPING THE AC SECRET

In March 1888, Professor William Anthony at Cornell University, was shown the AC machines under pledge of secrecy, as applications were still in the Patent Office. Nevertheless, he wrote excitedly to a friend:

> *I have seen an armature weighing 12 pounds running at 3,000 rpm, when one of the (ac) circuits was suddenly reversed, it reversed its rotation so suddenly I could hardly see what did it. In all this you understand there is no commutator. The armatures have no connection with anything outside ... It was a wonderful result to me in the form of motor I first described, there is absolutely nothing like a commutator, the two (ac) chasing each other round the field do it all. There is nothing to wear except the two bearings.*[3]

Professor Anthony and Thomas Martin were joined by Franklin Pope, editor of the *Electrical Engineer* and also a Westinghouse engineer and patent lawyer, in urging Tesla to present a paper to the AIEE. Martin would later write:

> *[I] had great difficulty in inducing Mr. Tesla to give the Institute any paper at all. Mr. Tesla was overworked and ill, and manifested the greatest reluctance to an exhibition of his motors, but his objections were at last overcome. The paper was written the night previous to the meeting, in pencil, very hastily, and under the pressure just mentioned.*[4]

Tesla made his own excuses to the AIEE, saying: "The notice was rather short, and I have not been able to treat the subject so extensively as I could have desired, my health not being in the best condition at present. I ask your kind indulgence, and I shall be very much gratified if the little I have done meets your approval."[5]

TESLA'S ELECTRIFYING DEMO

In his paper to the AIEE, Tesla sought to demonstrate the superiority of AC once and for all. He told the audience gathered in the lecture hall at Columbia College on 47th Street:

The subject which I now have the pleasure of bringing to your notice is a novel system of electric distribution and transmission of power by means of alternate currents, affording peculiar advantages, particularly in the way of motors, which I am confident will at once establish the superior adaptability of these currents to the transmission of power and will show that many results heretofore unattainable can be reached by their use; results which are very much desired in the practical operation of such systems and which cannot be accomplished by means of continuous currents. Before going into a detailed description of this system, I think it necessary to make a few remarks with reference to certain conditions existing in continuous current generators and motors, which, although generally known, are frequently disregarded.

In our dynamo machines, it is well known, we generate alternate currents which we direct by means of a commutator, a complicated device and, it may be justly said, the source of most of the troubles experienced in the operation of the machines. Now, the currents so directed cannot be utilized in the motor, but they must – again by means of a similar unreliable device – be reconverted into their original state of alternate currents. The function of the commutator is entirely external, and in no way does it affect the internal working of the machines. In reality, therefore, all machines are alternate current machines, the currents appearing as continuous only in the external circuit during their transit from generator to motor. In view simply of this fact, alternate currents would commend themselves as a more direct application of electrical energy, and the employment of continuous currents would only be justified if we had dynamos which would primarily generate, and motors which would be directly actuated by such currents.[6]

Nikola Tesla's original induction motor, 1888.

ELIHU THOMSON
(1853 – 1937)

Born in Manchester, England, Elihu Thomson immigrated to the US with his family when he was five. He was educated at the Central High School in Philadelphia, where he later taught chemistry and mechanics. Inspired by the work of Charles Brush, he and fellow teacher Edwin J. Houston (1847 – 1914) developed an arc lighting system that worked with AC.

In 1883, Thomson and Houston formed the Thomson-Houston Electric Company in Lynn, Massachusetts, and they went into the AC business. Westinghouse sued them for infringement of his Gaulard-Gibbs patents, until Thomson-Houston agreed to pay Westinghouse a royalty of $2 per horsepower on their transformers. By 1891, sales exceeded $10 million and the company boasted 666 central stations, compared to Westinghouse's 323 and Edison's 202. Thomson became president of the AIEE in 1889. Thomson's notable inventions include an arc lighting system, an automatically regulated three-coil dynamo, a magnetic lightning arrester, and a local power transformer.

Thomson-Houston Electric tried to stay out of the "battle of the currents," but a patents battle resulted in a merger with Edison General Electric in 1892 to form General Electric. Thomson continued research into x-rays and automobiles.

Tesla demonstrated his AC motors, showing that they could be reversed instantly. He provided precise calculations on how the speed and power of the motor could be determined, and he showed how his system could be married up to DC apparatus.

After the lecture, Professor Anthony said: "I confess that on first seeing the motors, the action seemed to me an exceedingly remarkable one." In his test, he had found Tesla's motors fifty to sixty percent more efficient than DC models.

POLARIZING OPINION

Then arc lighting pioneer, Elihu Thomson, stepped forward to say that he had already developed an AC motor.

I have been very much interested in the description given by Mr. Tesla of his new and admirable little motor. I have, as probably you may be aware, worked in somewhat similar directions, and towards the attainment of similar ends. The trials which I have made have been by the use of

> *a single alternating current circuit – not a double alternating circuit – a single circuit supplying a motor constructed to utilize the alternation and produce rotation.*[7]

However, Thomson's motor still used a commutator and was consequently inefficient. Tesla pointed out that his own motor with no commutator did not suffer this disadvantage, earning himself a life-long enemy in the process:

> *That peculiar form of motor represents the disadvantage that a pair of brushes must be employed to short circuit the armature coil.*[8]

> *You will see the advantage of this disposition of the closed circuit coil – that this action is always maintained at the maximum and it is indeed more perfect than if the polarities were shifted by means of a commutator.*[9]

MEETING MONSTROUS TERMS

Tesla's paper was printed in all the leading engineering journals. Westinghouse already knew of the induction motor as Franklin Pope, his employee, had visited Tesla's Liberty Street lab at Martin's behest. After studying

MEASURING CURRENT IN THE 1880s

As commercial use of electric energy spread in the 1880s, customers were initially charged monthly for the number of lamps they had. But as electrical energy found wider uses, an electric energy meter, similar to the already existing gas meter, was needed to properly bill customers for the amount they used. Edison developed an electrochemical metering system, which used an electrolytic cell to totalize current consumption. Periodically the plates were removed, weighed, and the customer billed. This was cumbersome and labor-intensive.

The "Reason" meter was developed in the UK. This comprised a vertical glass tube with a mercury reservoir at the top. As current was drawn from the supply, electrochemical action transferred the mercury to the bottom of the column. Once the mercury reservoir was empty, an open circuit was created. When the customer paid for a further supply of electricity, the supplier would unlock the meter from its mounting and invert it, tipping the mercury back into the reservoir.

In 1885 the Ferranti company devised a mercury motor meter with a register like that of a gas meter, which meant that the meter could be easily read by both the customer and the supplier's agent. The first accurate, recording electricity consumption meter was a DC meter by Dr. Hermann Aron, who had patented it in 1883. Hugo Hirst of the British General Electric Company introduced it commercially into Great Britain from 1888.

Unlike their AC counterparts, DC meters did not measure energy directly, but rather charge in ampere-hours, displayed on a series of dials. But as the voltage of the supply would remain more or less constant, this was proportional to actual energy consumed.

The first AC meter was made under the patent of Hungarian Ottó Bláthy and manufactured by Ganz. It measured energy consumption in kilowatt-hours and was shown at the Frankfurt Fair in the autumn of 1889. The first induction kilowatt-hour meter was marketed by the factory at the end of the same year. Around the same time, Elihu Thomson developed a recording watt-hour meter based on an ironless commutator motor. Overcoming the disadvantages of older electrochemical meters, it could operate on either alternating or direct current.

In 1894, at the Westinghouse Electric Corporation, Oliver Shallenberger applied the induction principle previously used in AC ampere-hour meters to produce a watt-hour meter. Purely electromechanical, it used an induction disk whose rotational speed was proportional to the power in the circuit. The induction meter worked on AC, however it eliminated the delicate and troublesome commutator of Thomson's design. Shallenberger's ill health prevented him refining his initial large and heavy design, though he did also develop a polyphase version.

Tesla's paper, Westinghouse dispatched Henry M. Byllesby to evaluate them.[10]

Visiting Liberty Street with Brown and Peck, Byllesby reported to his boss the meeting with Tesla, saying:

> *His description was not of a nature which I was enabled, entirely, to comprehend. However, I saw several points which I think are of interest. In the first place, as near as I can get it, the underlying principle of this motor is the principle which Mr. Shallenberger is at work at this present moment. The motors, as far as I can judge from the examination which I was enabled to make, are a success. They start from rest and the reversion of the direction of rotation can be suddenly accomplished without any short-circuiting.*[11]

Byllesby then inquired about the price of the patents, only to be told by Brown and Peck that they had already been offered $200,000 plus $2.50 per horsepower. If Westinghouse was to make a better bid, he had until the end of the week.[12] Byllesby wrote:

> *The terms, of course, are monstrous; and I so told them there was no possibility of our considering the matter seriously but that I would let them know before Friday ... In order to avoid giving the impression that the matter was one which excited my curiosity, I made my visit short.*[13]

KEEPING OPTIONS OPEN

Brown and Peck granted Westinghouse a $5,000 option for six weeks. At the time, Westinghouse's man in Europe was considering the AC motor patent of Galileo Ferraris, a professor in Turin. But while Ferraris's motor was little more than a toy to prove the principle, Tesla's system was a serious piece of machinery designed for industrial applications. Nevertheless, Westinghouse paid out $1,000 for Ferraris's patent.[14]

Meanwhile at Westinghouse, Oliver B. Shallenberger, who had by then invented the induction meter which measured the power

HOW AN INDUCTION MOTOR WORKS

An electrical motor works through the interaction of two sets of magnets – one stationary, the stator – and one able to move freely – the rotor, which is usually mounted on a shaft that can rotate. It is possible to create a primitive motor using permanent magnets, but they tend to lose their strength over time. But, in practical motors, electromagnets are used. Both the stator and rotor are essentially coils of wire around metal cores. When the electricity is switched on, the core becomes magnetized and magnetic attraction draws the north pole of the rotor toward the south pole of the stator. But when it gets there, it stops. To keep the motor turning, the polarity of the electromagnets has to keep switching. In early DC motors, the polarity of the rotor was switched by using a split-ring commutator to supply current.

But Tesla simplified matters when he realized that you did not need to supply electricity to the electromagnets or the rotor at all. The alternating current in the coils of the stator would create a magnetic field that would induce an electric current in the coils of the rotor. This would then create a magnetic field of its own, which would be of the opposite polarity to that in the stator, causing the motor to turn.

A dynamo or generator is simply a motor worked in reverse. A motor converts electrical energy into a mechanical turning force. When you supply a mechanical turning force to a dynamo, you generate electricity. Add in a transformer, which similarly works through induction, and Tesla had created a complete system of generating, transmitting, and utilizing power.

consumption of AC customers, was also working on an AC motor. William Stanley had also built an AC motor, but like Thomson's it depended on a commutator and brushes.[15]

On July 5, 1888, Westinghouse, his options running out, wrote:

> *I have been thinking over this motor question very considerably, and am of the opinion that if Tesla has a number of applications pending in the patent office, he will be able to cover broadly the apparatus that Shallenberger was experimenting with, and that Stanley thought he had invented. It is more than likely that he will be able to carry his date of invention back sufficient time to seriously interfere with Ferraris, and that our investment there will probably prove a bad one. If the Tesla patents are broad enough to control the alternating motor business, then the Westinghouse Electric Company cannot afford to have others own the patents.*

DOING THE WORK OF THE WORLD

In this light, the Ferraris patents were useless. Instead Westinghouse made a deal with Brown and Peck. He would pay $25,000 in cash and $50,000 in notes – payable in three installments – plus the $2.50 royalty per horsepower on every AC Tesla motor, with $5,000 minimum paid in royalties the first year, $10,000 the second, and $15,000 the third.

Although his lawyers had beaten down Brown and Peck from the original $200,000, Westinghouse wrote:

> *With reference to the Tesla motor patents, the price to be paid seems rather high when coupled with all of the other terms and conditions, but if it is the only practicable method for operating a motor by the alternating current, and if it is applicable to street car work, we can unquestionably easily get from the users of the apparatus whatever tax is put upon it by the inventors.*[16]

Tesla was delighted with the deal, finding, in Westinghouse, a kindred spirit. They were to have a long association. He said:

> *George Westinghouse was, in my opinion, the only man on the globe who could take my alternating current system under the circumstances then existing and win the battle against prejudice and money-power. He was a pioneer of imposing stature and one of the world's noblemen.*[17]

Both men were idealists in their way, believing that their inventions could change the world. Though others were profiting from his work, Tesla was more concerned with his AC system benefiting the greatest number of people.

"No more will men be slaves to hard tasks," he told Anthony Szigeti. "My motor will set them free, it will do the work of the world."[18]

MANIFESTING LATENT FORCE

In July 1888, Tesla left Manhattan for Pittsburgh, where he met Westinghouse for the first time. Tesla was full of admiration for him:

> *Even to a superficial observer, his latent force was manifest. A powerful frame, well-proportioned, with every joint in working order, an eye as clear as a crystal, a quick and springy step – he presented a rare example of health and strength. Like a lion in the forest, he breathed deep and with delight the smoky air of his factories.*[19]

Tesla toured Westinghouse's electrical shops and met the engineers. Then he returned briefly to New York to wrap up his affairs before starting work in Pittsburgh as a consultant with the Westinghouse Electric Company.

TESLA ON WESTINGHOUSE

When Westinghouse died in 1914, Tesla wrote:

> *I like to think of George Westinghouse as he appeared to me in 1888, when I saw him for the first time. The tremendous potential energy of the man had only in part taken kinetic form ... Though past forty then, he still had the enthusiasm of youth. Always smiling, affable and polite, he stood in marked contrast to the rough and ready men I met. Not one word which would have been objectionable, not a gesture which might have offended – one could imagine him as moving in the atmosphere of a court, so perfect was his bearing in manner and speech. And yet no fiercer adversary than Westinghouse could have been found when he was aroused. An athlete in ordinary life, he was transformed into a giant when confronted with difficulties which seemed insurmountable. He enjoyed the struggle and never lost confidence. When others would give up in despair he triumphed. Had he been transferred to another planet with everything against him he would have worked out his salvation. His equipment was such as to make him easily a position of captain among captains, leader among leaders. His was a wonderful career filled with remarkable achievements. He gave to the world a number of valuable inventions and improvements, created new industries, advanced the mechanical and electrical arts and improved in many ways the conditions of modern life. He was a great pioneer and builder whose work was of far-reaching effect on his time and whose name will live long in the memory of men.*[20]

f
A'
b
b
b
c
c'
e
h
f
f'
e'
h'
A
a

PART THREE

THE WAR OF THE CURRENTS

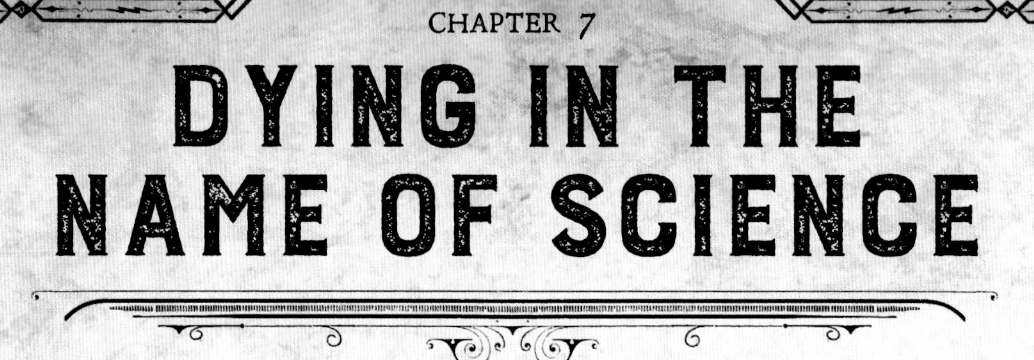

CHAPTER 7

DYING IN THE NAME OF SCIENCE

Although Edison had shamelessly exploited the electrocution of innocent members of the public to attack Westinghouse, there was growing public concern that electricity was not entirely safe. In the *New York Evening Post* of June 5, 1888, a letter appeared under the heading "Death in the Wires." The author was self-educated electrical engineer Harold P. Brown.

DEATH IN THE WIRES

The death of the poor boy Streiffer, who touched a straggling telegraph wire on East Broadway on April 15, and was instantly killed, is closely followed by the death of Mr. Witte in front of 200 Bowery and of William Murray at 616 Broadway on May 11, and any day may add new victims to the list.[1]

Brown went on to say that:

> *... several companies who have more regard for the almighty dollar than for the safety of the public, have adopted 'alternating' current for incandescent service. If the pulsating current [used for arc lights] is dangerous, then the alternating current can be described by no adjective less forcible than damnable. The only excuse for the use of the fatal alternating current is that it saves the company operating it from spending a larger sum of money for the heavier copper wires, which are required by the safe incandescent systems. That is, the public must submit to constant danger from sudden death, in order that a corporation may pay a little larger dividend.*[2]

Brown called for the outlawing of AC above 300 volts to "prevent the wholesale risk of human life."[3] Limiting AC to 300 volts would restrict its range and nullify its greatest advantage over Edison's DC system. It seems to be no coincidence that the *New York Evening Post* was owned by Henry Villard, a long-term investor in Edison's projects who went on to become president of Edison General Electric Company.

Three days after its publication, Brown insisted that the letter was read into the record at a meeting of the New York City Board of Electrical Control.

WESTINGHOUSE AND EDISON

While Westinghouse and Edison remained at loggerheads, the Pittsburgh pioneer tried to pour oil on troubled waters. On June 7, 1888, Westinghouse wrote a conciliatory letter, to Edison saying:

> *I have a lively recollection of the pains that you took to show me through your works at Menlo Park when I was in pursuit of a plant for my house, and before you were ready for business, and also of my meeting you once afterwards at Bergman's factory; and it would be a pleasure to me if you should find it convenient to make me a visit here in Pittsburgh when I will be glad to reciprocate the attention shown to me by you.*[4]

Knowing that Edison was in trouble, he even suggested a merger. Edison wrote back bluntly on June 12: "My laboratory work consumes the whole of my time and precludes my participation in directing the business policy ... Thanking you for your kind invitation to visit you in Pittsburgh."[5]

After this rebuff, Westinghouse adopted a very different tone when he wrote again on July 3. This time he enclosed a confidential memorandum that Edison had sent out to his company. It said: "If the Westinghouse Company, as they claim, have a system so superior and so much cheaper than Edison, and if, as they pretend, the Edison Company's patents are valueless, why do they desire to bring about a combination with the Edison Company."[6]

Topsy the Elephant's death by electrocution was a horrifying publicity event staged for the opening of a new amusement park on Coney Island at Luna Park on January 4, 1903. Organized by park owners, Frederick Thompson and Elmer Dundy, the cruel killing was recorded by an Edison film crew, who were among the press that day, and a short kinetoscope film was released called *Electrocuting an Elephant*. In popular culture, the film has become erroneously linked to Harold Brown's anti-AC demonstrations. The reality is that the War of the Currents had been 10 years earlier and the events at Luna Park had nothing to do with Edison. (Above) Topsy, surrounded by press photographers and on-lookers, in the middle of Luna Park, which was still under construction.

WESTINGHOUSE FIGHTS BACK

In mid July, Westinghouse replied to Brown's attack at the New York City Board of Electrical Control, saying that creating "this enormous business within so short a time ... that it has been considered inexpedient, heretofore, to take any notice of, or make any reply to, the criticism and attacks of some of the opposition electric lighting companies."[7]

He said that the "method of attack" used against him "has been more unmanly, discreditable and untruthful than any competition which has ever come to my knowledge." And he went on to defend his AC system's safety record:

> *...we have yet to receive knowledge of a single case of fire of any description from the use of our system. Of the 125 central stations of the leading direct current company [Edison's] there were numerous cases of fire, in three of which cases the central station itself was entirely destroyed, the most recent being the destruction of the Boston station; while among the almost innumerable fires caused by this system, among the users, may be mentioned the total destruction of a large theater at Philadelphia.*[8]

Westinghouse also bombarded the New York City Board of Electrical Control with affidavits. One came from a William L. Wright, an employee of the Keystone Electric Light and Power Company. He had been working in a damp cellar where the lighting had previously been supplied by a 200-volt three-wire DC system.

> *The wiring there had been changed and connected to the 1,000 alternating current, and forgetting that this current would be on the socket, I took hold of the socket while standing on the wet ground, when I received a shock which threw me on my face with my hand underneath me, and still grasping the socket ... When I came to my senses, I was sitting in the cellar held up by two of the men. In the meantime, an ambulance had been called, but I refused to go in it, as I did not feel in any way injured except that my hand had been badly burned. I went down to the electric light station and waited there for fifteen or twenty minutes to receive my money; it being pay-day, and then went home.*[9]

Company regulations insisted that he visit the doctor.

> *These burns healed very slowly; but I have not felt in any way any of the after effects from this shock, such as are usually felt from high tension direct current machines ... and I feel sure that had I received this kind of a shock from a direct current machine of any of the ordinary types ... it would have been fatal.*[10]

Harold Brown (1857 – 1944).

HAROLD BROWN AND "DAMNABLE" AC

In the face of Westinghouse's stout defense, Harold Brown was determined to prove his point. He needed a powerful ally, so he called on Edison, who he claimed not to know – "To my surprise, Mr. Edison at once invited me to make experiments in his private laboratory, and placed all necessary apparatus at my disposal." [11]

Not only that, but Edison assigned his chief electrician, Arthur Kennelly, to work with Brown, while promising to take a special interest himself. In Edison's new laboratories in West Orange, Brown set about making a series of experiments to show that AC was indeed "damnable." Meanwhile Edison wrote to the American Society for the Prevention of Cruelty to Animals, asking for some good-sized dogs to experiment on.[12]

DYING FOR SCIENCE'S SAKE

On July 30, 1888, seventy-five electricians were invited to attend a demonstration by Harold Brown at the private laboratory of Professor Thomas Chandler at the Columbia College School of Mines. Brown told the audience that he represented no company and no financial or commercial interest. He was there, he said, on behalf of the many workmen who had been accidentally electrocuted.[13] Not only had they lost their lives, but their employers had denigrated their memory by saying that they had caused their own deaths by their "foolish carelessness."

Brown maintained that it was the fault of AC. He said he had proved by repeated experiments that a living creature could stand shocks from a continuous current much better than from an alternating current. He had applied 1,410 volts of DC to a dog without fatal results, but had repeatedly killed dogs with 500 to 800 volts of alternating current.[14]

To assure the audience that the demonstration was to be performed honestly, he would be assisted by Arthur Kennelly, Dr. Frederick Peterson, a physician who specialized in the medical uses of electricity, Dr. Schuyler S. Wheeler of the New York Board of Electrical Control, and T. Carpenter Smith, a leading critic of Brown.[15]

Brown then brought out a Newfoundland mix named Dash weighing 76 pounds (34 kilograms). The dog was muzzled and tied down inside a wooden cage with copper wires wound round the bars. The audience was told that, while the dog looked friendly, he was a "desperate cur" who had already killed two people.

Then electrical contacts were fitted to Dash's right fore-leg and left hind-leg. They were wrapped in wet cotton wadding and bound with copper wire. The dog's resistance was then measured at 15,300 ohms.[16]

What happened next was reported under the headline: "Died for Science's Sake – A Dog Killed With The Electric Current" in *The New York Times* the following day:

> *Mr. Brown announced that he would first try the continuous current at a force of* 300 *volts. When the shock came the dog yelped and then subsided. A relay has been attached to the apparatus, which shut off the current almost as soon as it was applied. When the dog got* 400 *volts he struggled considerably, still yelping. At* 700 *volts he broke his muzzle and nearly freed himself. He was tied again. At* 1,100 *volts his body contorted with the pain of the brutal experiment. His resistance to the current then dropped from* 15,000 *to about* 2,500 *ohms.*
>
> *"He will be less trouble," said Mr. Brown, "when we try the alternating current. As these gentlemen say, we shall make him feel better."*
>
> *It was proposed the dog be put out of his misery at once. This was done, with an alternating current of* 330 *volts killing the beast.*[17]

When Brown brought out another dog, an agent from the American Society for the Prevention of Cruelty to Animals stepped in.

He said that "if it was necessary to torture animals in the interests of science it should be done by the colleges and institutions not by rival inventors."[18]

Meanwhile, the assembled electricians said that the dog had been weakened by the DC current before the AC was applied. But Brown insisted that he had enough dogs to prove his point to the most skeptical of observers.

"The only places where an alternating current ought to be used were the dog pound, the slaughter house and the State prison,"[19] he said.

THE HORROR SHOW CONTINUES

Brown was not to be stopped. Four days later, he performed more demonstrations at the Columbia College School of Mines, this time under the auspice of Dr. Charles F. Roberts, assistant professor of physiology at Bellevue Hospital Medical College, and Cyrus Edson of the Board of Health. There were nearly eight hundred in the audience.[20]

With no representative of Westinghouse present, Brown electrocuted three dogs, only using alternating current. The first had a resistance of 14,000 ohms and died after five seconds of 272 volts. The second's resistance was 8,000 ohms and it died after five seconds at 340 volts.[21]

Next came a Newfoundland-setter mix. The *New York Morning Sun* said that this "big black beast" looked "bright and intelligent," and appeared friendly. Its resistance was 30,000 ohms. Brown administered 220 volts of AC for five seconds. The dog was paralyzed while the current was on. But when the current was switched off, it struggled in agony. Brown then gave it thirty seconds of 220 volts. It remained conscious. After forty-five seconds, its breathing became labored. After sixty, it appeared to be recovering, so Brown wound the voltage up to 234 volts. It stood paralyzed for another thirty seconds, then collapsed.[22]

The current was switched off, but the dog's heart continued beating for a further two minutes. Brown and the doctors present maintained that it had felt no pain. However, like the other dogs, it had lost control of its bowels and bladder. The smell of feces and urine were now so overwhelming that the demonstration had to be ended.[23]

The *Electrician* wrote: "If this sort of thing goes on, with the accidental killing of men and the experimental killings of dogs, the public will soon become as familiar with the idea that electricity is death as with the superstition that it is life."[24]

Brown was well satisfied though. He wrote to Kennelly on August 5, saying: "It is certain that yesterday's work will get a law passed by the legislature in the fall, limiting the voltage of alternating currents to 300 volts."

But critics were not convinced. There is an old saying among electricians: "Its volts that jolts, but mils that kills." In other words, if the voltage is high but the current – measured in milliamperes – is low it will produce an unpleasant sensation but little damage. It is the current that causes injury.

REVEALING THE DECEPTION

Writing in the *Electrical Engineer* in September 1888, Dr. P.H. Van der Weyde accused Harold Brown of deception. The dynamos he had used had been supplied by Edison and, while only one had been used to administer DC, two had been used to supply AC – giving double the current. Van der Weyde compared Brown to those people who had warned that the introduction of gas lighting would blow up cities, and railroad trains would result in the wholesale slaughter to those who tempted Providence by traveling at the outrageous speed of twenty miles an hour.[25]

In the market place, Brown's experiments had little effect. In the month of October 1888 alone, Westinghouse took orders for 48,000 lights powered by AC central stations. The company had then sold more central stations than all the DC companies combined.[26]

CHAPTER 8

FACING THE ELECTRIC CHAIR

In 1888, the New York State Legislature adopted electrocution as the official method of capital punishment. Chairing the committee of New York's Medico-Legal Society investigating how this was to be done, was none other than Dr. Frederick Peterson, fresh from assisting Harold Brown with his dog-slaying experiments.[1]

Peterson then joined Brown in Edison's West Orange lab where they carried out more experiments. On November 15, they told a meeting of the Medico-Legal Society that both AC and DC would do the job, but that they favored AC.

The Society was to make its decision at a meeting on December 12, but to help them make up their minds Brown and Peterson would put on more demonstrations on December 5. This time they would use larger animals, nearer to the size of a grown man. Members of the Medico-Legal Society and the New York State Death Commission would join reporters and physicians in West Orange. Edison would also be there.[2]

DEATH-CURRENT EXPERIMENTS

The first victim was a calf bought from the local butcher. It weighed 124 pounds. Hair was clipped from between its eyes and between its shoulders. Electrodes were attached using sponges soaked in a conducing solution of zinc sulfate. The dynamo was cranked up to 1,100 volts and the circuit closed. According to Arthur Kennelly's notes: "The animal fell uninjured."[3] It was thought that the problem lay with the transformer, so they disconnected it and wired the calf up directly to the dynamo. With the voltage run up to 770, the circuit was closed for eight seconds.

"Death was instantaneous," Brown claimed in his published report.[4]

A second calf, weighing 145 pounds, died after receiving 750 volts for five seconds. Then a horse weighing 1,230 pounds was led in.[5]

Edison had suggested that a convicted criminal could be electrocuted by placing an electrode on each hand, so that the current passed through his chest. To test this, Brown and Kennelly attached electrodes to the animal's forelegs. Then they gave it a shock of 600 volts for five seconds. It did not die, so they gave it a second shock for 15 seconds. Again it survived. Eventually 700 volts for 25 seconds killed it.[6]

The following morning, *The New York Times* solemnly reported:

> *The experiments proved the alternating current to be the most deadly force known to science, and that less than half the pressure used in this city* [1,500 *to* 2,000 *volts*] *for electric lighting by this system is sufficient to cause instant death. After Jan.* 1, *the alternating current will undoubtedly drive the hangman out of business in this State.*[7]

Kennelly's notes indicated that the calves and horses had suffered, just as the dogs had. But Brown insisted in his report published in the *Electrical World* that the deaths of all three animals were, "as in the other cases ... instantaneous and painless." He concluded that "the alternative current is the best adapted for electrical executions."

On December 12, the Medico-Legal Society unanimously recommended that condemned prisoners could be given a shock of alternating current between 1,000 to 1,500 volts for 15 to 30 seconds.[8]

Harold Brown demonstrates the killing power of AC by electrocuting a horse at Thomas Edison's West Orange laboratory, 1888.

The *Daily Tribune* concluded that Brown's experiments "showed that less than half the pressure used for electric lights in our streets is sufficient to produce instant death. Evidently the danger from electric light wires has not been over-estimated."[9]

KILLING THE RUMORS

The following day, Westinghouse's response to Brown's demonstrations appeared in *The New York Times.* The public, he said, were being misled and the public interest injured. Brown was in the pay of the Edison Electric Company. While Brown claimed that a shock of 300 volts of AC would kill, Westinghouse denied it:

> *It had been found that pressures exceeding 1,000 volts can be withstood by persons of ordinary health without experiencing any permanent inconvenience. Further, the alternating current is less dangerous to life, from the fact that the momentary reversal of direction prevents decomposition of tissues, and injury can only result from the general effects of the shock ...*[10]

Westinghouse said a large number of people who had received a 1,000-volt shock from alternating current could be produced, including a wireman who had held onto the live cable for three minutes and was able to go on with his work after a short period. His AC system was now being used in London "where the laws with reference to the distribution of electricity are more carefully scrutinized than they ever have been in this country." What's more, AC was actually safer. "It not only permits the use of the current of 1,000 volts for street mains, but requires its conversion into currents of 50 volts or less for house-wiring," Westinghouse wrote.

> *The converters are so constructed that the primary or street current can never by any possibility enter the house. With the Edison*

system the pressure is about 230 volts, and while no person coming in contact with the alternating current as used for domestic lighting would be aware of its presence, with the Edison system the shock would be painful if not absolutely dangerous, if the person were at all delicate.[11]

THROWING DOWN THE CHALLENGE

Brown responded in the pages of the *New York Tribune* on December 18, citing Westinghouse's "pecuniary interests ... [in] the death-dealing alternating current" that had "crippled, paralyzed or otherwise injured for life a number of men."[12]

Mr. Westinghouse asserts that the alternating is less dangerous than the continuous current, but has failed to prove it. I therefore challenge Mr. Westinghouse to meet me in the presence of competent electrical experts, and take through his body the alternating current, while I take through mine a continuous current. The alternating current must not have less than 300 alternations per second (as recommended by the Medico-Legal Society) ... I will warn Mr. Westinghouse, however, that 160 volts alternating current for five seconds has proved fatal in my experiments, and that several men have been killed by the low-tension Jablochkoff [arc lamp] *alternating current.*[13]

Westinghouse did not rise to the challenge ... but Nikola Tesla would.

EXECUTION BY ELECTRICITY

With AC now being dubbed the "executioner's current," Brown was hired by the New York State prisons to design their electrocution equipment. Naturally, he would use Westinghouse generators to produce the electricity. The *Electrical Review* detailed Edison's recommended method:

He proposes to manacle the wrists, with chain connections, place ... the culprit's hand in a jar of water diluted with caustic potash and connect therein ... to a thousand volts of alternating current ... place the black cap on the condemned, and at a proper time close the circuit. The shock passes through both arms, the heart and the base of the brain, and death is instantaneous and painless.[14]

Soon they had their guinea pig in the shape of William Kemmler, a thirty-year-old alcoholic who had killed his common-law wife, Tillie Ziegler, in Buffalo. Suspecting she was having an affair with his business partner and lodger, he delivered twenty-six blows to her head, neck, and chest with an axe, in front of her four-year-old daughter, Ella. There was no doubt of his guilt. "I had to do it," he told a neighbor. "There was no help for it. Either one of us had to die. I'll take the rope for it."

After he was arrested, he was asked how he had killed Tillie – "With a hatchet," he said, "I wanted to kill her, and the sooner I hang, the better it will be."[15]

He made a statement and there was no attempt to challenge the facts of the case in court. The plea of "alcoholic insanity" was the defense attorney's one desperate chance for his client to escape the death penalty. He was found guilty of murder in the first degree. On Tuesday, May 14, 1889, he was the first person to be sentenced to death under the new Electrical Execution Act. Pronouncing sentence, Judge Henry Childs said:

The sentence is that for the crime of murder for which you stand convicted that within the week commencing Monday, June 24, 1889, within the walls of Auburn State prison or within the yard or enclosure thereof, you suffer the death punishment by being executed by electricity, as provided by the code of criminal procedure of the State of New York, and that you be removed to and kept in confinement in Auburn State prison until that time. May God have mercy on your soul.

Kemmler's lawyers appealed on the grounds that electrocution was a cruel and unusual punishment. His defense team was now joined by noted orator William Bourke Cockran, who was paid by Westinghouse.

UNCOVERING THE EVIDENCE

Cockran sought a writ of habeas corpus "to establish facts upon which the court can pass, as to the character of the penalty." [16] The New York Supreme Court granted a stay of execution pending the results of an evidentiary hearing. The first witness to be called was Harold Brown and Cockran set about trashing his credentials.

In a fierce cross-examination, Cockran established that Brown had no education in engineering, was not a member of the Institution of Electrical Engineers, and had no professional standing.[17] Nor was Brown a physician or surgeon. He had no medical knowledge of the precise effects of electricity on the human body.[18]

Cockran then turned his attention to the electric chair itself. Brown had build three – one each for the state prisons at Ossinging (Sing Sing), Auburn, and Clinton. Each was furnished with a 650-light Westinghouse alternating-current dynamo.[19] He asked for a drawing of the apparatus, which was easy to characterize as a device straight from a medieval torture chamber.[20]

Despite his denials, it was easy to show that Brown was a close associate of Edison and that the purpose of his experiments was not to discover a quick and painless method of execution, but rather to demonstrate that direct current was "comparatively harmless" while alternating current was "extremely dangerous." [21] However, his only knowledge of lethal electric shocks to human beings came from accidental deaths.

Arthur Kennelly (1861 – 1939).

THOMAS EDISON TESTIFIES

The hearings moved from New York City out to Edison's laboratories in West Orange and Thomas Edison was called to testify on behalf of the People. He, naturally, maintained that AC was more likely to produce a lethal shock, though he denied any animosity toward Westinghouse, even though they were locked in a long-running legal battle over patent violations.[22]

Arthur Kennelly followed Edison onto the witness stand, and Cockran put it to him that the purpose of the animal experiments was to conceal the fact that direct current could kill. Kennelly, too, was forced to admit that no one really knew the effect of electricity on the human body, and even confessed that he could not be sure that some of the animals he had helped electrocute were actually dead before the autopsy was performed.[23]

The hearings moved on to Buffalo, where testimony was heard that dogs thought to

WILLIAM BOURKE COCKRAN
(1854 – 1923)

Born in County Sligo, Ireland, and educated in France, Cockran immigrated to the US when he was seventeen where he studied law and was admitted to the bar in 1876. He came to national attention with a rousing speech at the Democratic National Convention in 1884.

Said to have been a one-time lover of Winston Churchill's Brooklyn-born mother Jennie Jerome, he introduced the young Winston to American society on his first visit to the US at the age of twenty. Churchill once confided to Adlai Stevenson that he based his oratorical style on Cockran's.

Cockran was not a civic-minded attorney who did pro-bono work for indigent clients. Although he dodged the question, the newspapers all asserted that he was employed by Westinghouse in the Kemmler case for a reputed $100,000. Cockran himself claimed that he took the case for his love of dogs.

An "intimate friend" of Cockran told the *Electrical Review* that "Cockran has a remarkably fine dog and he is very fond of him. His wife, too, thinks a great deal of him [the dog]. When the experiments were going forward at Edison's laboratory and at the Columbia College School of Mines, Cockran and his wife used to read of the suffering of the poor dogs which were made the victims of science. One night Mrs. Cockran had finished reading a peculiarly graphic account of a dog's suffering, when she said: 'Just think how terrible it would be if they would treat our dog like that.' Cockran sprang up and exclaimed, 'The law providing for execution by electricity is unconstitutional. I'll beat it if I can.' When Kemmler was sentenced he immediately took charge of the case, although he knew he wouldn't get a cent."[24]

The *New York Tribune* was not fooled. After quoting Cockran as professing "his love for humanity and his desire to prevent an inhumane execution," which he said the electric chair would cause, the paper went on to report: "It leaked out that he was in the pay of the Westinghouse Company."[25]

Cockran went on to become a Congressman.

have been electrocuted had been revived and that autopsies on people accidentally electrocuted showed no discernible cause of death. Finally, the warden of Auburn prison, Charles Durston, testified that the electric chair at the prison was not going to be tested before being used on Kemmler.

"So that, in point of fact, your first test will be on Kemmler?" Cockran asked.

"Yes, sir," said Durston.[26]

I AM READY TO DIE

While the report of the evidentiary hearings was still at the printer, some of Brown's private letters were published in the *New York Sun*. These showed that he colluded with Edison to smear Westinghouse in their campaign to prove that AC was dangerous.[27] Nevertheless, Judge S. Edwin Day of Cayuga County Court, found that the use of the electric chair was not cruel and unusual,

The execution of William Kemmler, August 6, 1890

as the Electrical Execution Act had only become law after considerable deliberation by the state legislature. The burden of proof to establish that the law was unconstitutional must be borne by the prisoner. [28]

It had not been shown, beyond reasonable doubt, that electrocution was slower and more painful than hanging. The constitutionality "of an act is to be presumed," the judge said, and the responsibility of showing that an act is unconstitutional should not fall on the shoulders of a single judge.[29]

Westinghouse's attorneys immediately appealed to the New York Supreme Court. The Court found that while execution by electrocution was unusual, it was not cruel as it did not involve torture or a lingering death. Indeed, it was assumed to be instantaneous and painless. The fact that no one knew exactly how electricity caused death was irrelevant, as no one knew precisely how gravity worked, though it was used in hanging.[30]

More appeals were heard, but on March 31, 1890, when the sentence of death in the electric chair was confirmed, Kemmler told the court: "I am ready to die by electricity. I am guilty and must be punished. I am ready to die. I am glad I am not going to be hung. I think it is much better to die by electricity than it is by hanging. It will not give me any pain." [31]

The continued delays were blamed on Westinghouse. The *New York Tribune* harangued him in an editorial headlined: "Who is Kemmler's friend?" [32] Westinghouse denied any involvement in the campaign to save Kemmler until two days after his execution. His objection, he said, was simply because of the use of a Westinghouse dynamo which was not designed or adapted to be used as a means of execution.[33]

This was not unreasonable, as gloating Edison Electric executives were speaking of criminals being "Westinghoused," [34] in the hope that their rival's name would be immortalized in the same way that Dr. Guillotine's had been in France.

KEMMLER GOES TO THE CHAIR

The execution of William Kemmler went ahead on August 6, 1890. It was neither instantaneous nor painless. Nor was it dignified. After the hair had been shaved from the top of his head, a hole was cut in the seat of his pants so that the electric chair's electrodes would make contact. But once Kemmler was sitting in the chair, he had to stand up again so that the tail of his shirt could be cut away by the Warden, Charles Durston, with a pocket knife.[35]

The chair had been moved to a dingy basement room, illuminated by gaslight. Following the hearings, Warden Durston had authorized that the electric chair be tested beforehand by the killing of a calf. When Kemmler was ushered in at 6.32 a.m., twenty-five official witnesses were waiting. These included leading physicians and lawyers, along with two selected reporters.

Kemmler told them: "Gentlemen, I wish you all good luck. I believe I am going to a good place, and I am ready to go. I want only to say that a great deal has been said about me that is untrue. I am bad enough. It is cruel to make me out worse." [36] As Kemmler settled back in the chair, Durston attached the rear electrode with its suction cup and wet sponges.

"Now take your time and do it all right, Warden," said Kemmler. "There is no rush. I don't want to take any chances on this thing, you know." [37]

Durston adjusted the leather straps across Kemmler's forehead and chin, partially obscuring his features. When Durston stepped back, Kemmler shook his head and said: "Warden, just make that a little tighter. We want everything all right, you know." [38]

After tightening the head straps, Durston buckled the straps around his body, arms, and legs – eleven straps in all. Kemmler tested each one to see that it was tight enough.

THROWING THE SWITCH

"Goodbye, William,"[39] said Durston. This was the signal to the "State Electrician" – or executioner – Edwin F. Davis, to throw the switch. The *New York Times* reported the horrific scene:

> *Simultaneously with the click of the lever the body of the man in the chair straightened. Every muscle of it seemed to be drawn to the highest tension. It seemed as though it might be thrown across the chamber were it not for the straps which held it. The body was as rigid as though it had been cast in bronze, save for the index finger of the right hand, which closed up so tightly that the nail penetrated the flesh on the first joint and the blood trickled out on the arm of the chair.*[40]

Two doctors watched closely, while a third held a stopwatch. Kemmler was motionless and ashen. After seventeen seconds, one of the attending physicians, Dr. Edward Charles Spitzka, said: "He is dead."[41]

At 6.43 a.m., the current was switched off. The witnesses breathed a sigh of relief and they turned their heads away as Durston began to remove the electrode from Kemmler's scalp.[42]

MORE CIVILIZED WAY TO DIE

After a brief examination of the corpse, the chair's inventor, Dr. Alfred Southwick, turned to the witnesses and said: "This is the culmination of ten years work and study. We live in a higher civilization from this day."

Then they noticed that blood was still trickling from his finger wound. His heart was still pumping.

"Great God!" said someone. "He is alive."

"Turn on the current," said another.[43]

"To the horror of all present, the chest began to heave, foam issued from the mouth, and the man gave every evidence of reviving,"[44] wrote the *Electrical Review*.

"For God's sake, kill him and have it over," yelled a witness.[45]

The reporter from the Associated Press fainted, while the district attorney groaned and rushed from the room.[46]

Durston began reattaching the electrode. By then everyone could see Kemmler's chest rising and falling. He was breathing. Durston gave the signal for the current to be turned back on. Blood began to appear on Kemmler's face as small blood vessels under the skin ruptured. Then a terrible smell permeated the room as his hair and flesh began singeing.[47]

"The stench was unbearable," said *The New York Times*. "How long this second execution lasted ... is not really known by anybody. Those who held watches were too much horrified to follow them."[48]

The current was turned off again at 6.51 a.m. This time Kemmler was surely dead, though Dr. Spitzka insisted that they leave the body for three hours before they begin the autopsy, so that no one could say that he died under the knife, rather than from the electric current.[49]

ROASTED ALIVE

One of the doctors present told *The New York Times* he would rather have seen ten hangings.[50] Westinghouse said they would have done better with an axe, while Tesla said that the electric chair was "an apparatus monstrously unsuitable, for the poor wretches are not dispatched in a merciful manner but literally roasted alive. To the observer their sufferings seem to be of short duration; it must be borne in mind, though, that an individual under such conditions, while wholly bereft of the consciousness of the lapse of time, retains a keen sense of pain, and a minute of agony is equivalent to that through all eternity."[51]

Edison blamed the doctors. They had applied the current to the top of the head, though hair was not conductive, and they had not put his hand in a jar of water. However, next time they would get it right, he said.[52] Brown was notable by his absence.[53] He explained that he had been hired to supply three electric chairs and his contract expired on May 1, 1890.

FAR WORSE THAN HANGING

The New York Times

Auburn, N.Y. August 6, 1890.

A sacrifice to the whims and theories of the coterie of cranks and politicians who induced the Legislature of this State to pass a law supplanting hanging by electrical execution was offered today in the person of William Kemmler, the Buffalo murderer. He died this morning under the most revolting circumstances, and with his death there was placed to the discredit of the State of New York, an execution that was a disgrace to civilization.

Probably no convicted murderer of modern times had been made to suffer as Kemmler suffered. Unfortunate enough to be the first man convicted after the passage of the new execution law, his life has been used as the bone of contention between the alleged humanitarians who support the law, on one side, and the electric-light interests, who hated to see the commodity in which they deal reduced to such a use as that. For fifteen months they have been fighting as to whether he should be killed or not, and the question has been dragged through every court. He has been sentenced and resentenced to death, only to be dragged back from the abyss by some intricacy of the law ...[54]

The electric chair, as reconstructed for the movie *The Green Mile* (1999).

CHAPTER 9

SHOOTING SPARKS & ELECTRIC FLAMES

Tesla soon found himself at loggerheads with Westinghouse, who wanted his induction motors for the lucrative streetcar business. However, the Serb's polyphase motors did not sit well with Westinghouse's single-phase AC system. Tesla's motors were designed to work at sixty cycles per second, which he worked out was the optimum frequency. Westinghouse's power stations had been set up to run lighting systems. They used 133 cycles per second to minimize flicker.

Tesla set about trying to adapt his motors. His assistant was Charles Scott, who would later become professor of engineering at Yale, but back then, had only just learned that there was such a thing as alternating current, after reading an article written by William Stanley in *Electrical World*.

"I had ... graduated from college two years earlier, and I wondered why I had not heard of such things from my professors,"[1] said Scott.

Sarah Bernhardt (1844 – 1923).

TESLA GOES TRAVELING

After just over a year with Westinghouse, Tesla left Pittsburgh and went to Paris as part of a delegation from the AIEE to the *Congrès International de Electriciens*. This was being held in conjunction with the Universal Exposition of 1889 where the Eiffel Tower made its first appearance. He met Norwegian scientist Vilhelm Bjerknes (1862 – 1951) whose study of electrical resonance was vital in the development of radio.

He also ran across the famous French actress Sarah Bernhardt (1844 – 1923). Legend has it that she dropped her lace handkerchief and Tesla handed it back to her without even raising his eyes. The *Electrical Review* remarked that he may be "invulnerable to Cupid's shafts."[2] However, Tesla said, much later, that Bernhardt had dropped a scarf, not a handkerchief, and he did not return it. He kept it, without washing it, for the rest of his life.

Tesla paid a short visit to his family, then returned to New York, where he opened a laboratory on Grand Street and began his high-frequency research.

"I was not free in Pittsburgh," he said. "I was dependent and could not work ... When

I [left] that situation, ideas and inventions rushed through my brain like Niagara."[3]

What he developed was a new invention that would make another dent in the Edison empire. These would be light bulbs that needed no filaments and no wires. They were filled with gases that glowed in a highly charged, high-frequency AC atmosphere.

ROLLING OUT THE RED CARPET

Edison also attended the Universal Exposition of 1889 in Paris. He had been given a one-acre site to display his inventions and the latest – the phonograph – was causing a sensation. With his new wife, twenty-four-year-old Mina, Edison had lunch with Alexandre Gustave Eiffel in his apartment at the top of the tower. The composer Charles-François Gounod (1818 – 93) played for them. Edison also visited Louis Pasteur (1822 – 95) in his laboratory and received the Grand Cross of the *Légion d'Honneur* for his achievements.

He went on to Germany to meet Hermann von Helmholtz, who developed the mathematics of electrodynamics, and the electrical engineer Werner von Siemens, traveling with them to a meeting of the German Association of Science at Heidelberg.

Moving on to London, he visited Sir John Pender, who he described as "the master of the cable system of the world at the time."[4] Pender introduced Edison to the backers of Sebastian Ziani de Ferranti who was building an AC power station in Deptford, south London, that was able to transmit electricity at 11,000 volts to central London, seven miles away. Edison said:

> *They insisted I should give them some expression of my views. While I gave them my opinion, it was reluctantly; I did not want to do so. I thought that commercially the thing was too ambitious, that Ferranti's ideas were too big, just then; that he ought to have started a little smaller until he was sure. I understand that this installation was not commercially successful, as there were a great many troubles. But Ferranti had good ideas, and he was no small man.*[5]

Again Tesla's opinion greatly diverged from that of Edison. He would promise to "place a 100,000 horsepower on a wire and send it 450 miles in one direction to New York City, the metropolis of the East, and 500 miles in the other direction to Chicago, the metropolis of the West."[6]

Nikola Tesla holding a glowing, gas-filled light bulb.

(Upper) French engineer Alexandre Gustave Eiffel (1832–1923), poses high on the steps of the completed Eiffel Tower, 1889. (Lower) At the base of the tower, men and machinery raised the elevator.

The Eiffel Tower at The Paris Exposition, 1889.

HERMANN VON HELMHOLTZ
(1821 – 94)

Like many scientists of the day, Helmholtz worked in various fields, including physiology, optics, meteorology, hydrodynamics, and the philosophy of science. He is best known for the Law of Conservation of Energy, first published in 1847.

Born in Prussia, he was the son of a professor of philosophy and literature at the Potsdam Gymnasium. His mother was a descendent of William Penn, founder of Pennsylvania. The young Helmholtz turned his back on the ideas of natural philosophy and sought scientific insight from empirical observations of the real world.

He studied to become an army doctor in Berlin, but also attended lectures on physics and applied himself to mathematics. In his barracks in Potsdam, he set up a laboratory, then went on to become an academic, first in the field of physiology, then physics.

Between 1869 and 1871, he studied electrical oscillation, and he noted the oscillations of electricity in a coil when it was connected to a Leyden jar. At a lecture delivered in London in honor of Michael Faraday in 1881, he proposed that electricity was made up of particles, which soon became known as electrons. He sought to measure the speed of electromagnetic induction, but left the determination of the length of electric waves to his favorite pupil, Heinrich Hertz.

WERNER VON SIEMENS
(1816 – 92)

After attending grammar school in Lübeck, Siemens joined the Prussian School of Artillery where he was trained as an engineer. He was also interested in chemistry, carrying out experiments in his cell after being jailed for acting as a second in a duel between fellow officers.

In 1841, he joined the artillery workshop in Berlin, where he developed an electroplating process. He then began working on improvements to the electric telegraph, invented by Sir Charles Wheatstone (1802 – 75) in 1837. After laying a number of government lines, Siemens quit the army to set up as a manufacturer of telegraph systems with Johann Georg Halske (1814 – 90). With his brother Carl (1829 – 1906), Siemens set up operations in London, Paris, Vienna, and St. Petersburg.

His use of gutta-percha – latex from trees, also used in early golf balls – as insulation made the first underground and submarine cables. Siemens & Halske extended the telegraph system across the Mediterranean and from Europe to India.

In 1866, Siemens invented the self-excited generator, where some of the electricity generated is used to power the electromagnetic coils. Other inventions include the first electric-powered elevators, a prototype trolleybus, and the "dynamic" or moving-coil transducer that developed into the loudspeaker. His company also produced the tubes used by Wilhelm Conrad Röntgen (1845 – 1923) to investigate X-rays.

EDISON AT THE PINNACLE

While Edison had been away, Henry Villard (1835 – 1900), president of Edison Electric Light, had reorganized Edison's various companies into Edison General Electric, capitalized at $12 million. This gave Edison a windfall of $1.75 million.

"I have been under a desperate strain for money for twenty-two years," said Edison, "and when I sold out, one of the greatest inducements was the sum of cash received, so as to free my mind from financial stress, and thus enable me to go ahead in the technical field."[7]

President of the new conglomerate Edward H. Johnson (1846 – 1917) said: "We shall speedily have the biggest Edison organization in the world with abundant capital after which it will be goodbye to Westinghouse et al."[8]

The corporation then employed three thousand men, turned over $7 million a year, and made an annual profit of $700,000.[9]

STEPPING OUT TESLA-STYLE

Tesla was also riding high when he arrived back in New York that fall. He had outfitted himself with the latest styles from Paris. He wore soft-leather shoes and spats, and carried a cane. His silk handkerchiefs and kid gloves were thrown away after a weeks' use. Since his stay in Pittsburgh, he had developed a taste for luxury hotels and moved into the Astor House, dining nightly in Delmonico's.[10]

He insisted on being seated at a table that was not to be used by other customers. He needed a fresh tablecloth with every meal and two dozen napkins. The silverware had to be sterilized before it left the kitchen. Tesla would then pick up each item with one napkin, and polish it with another. Then he would drop both napkins on the floor before attending to the next item of cutlery. And if a fly alighted on the table, he would insist that everything was removed from the table and the meal would start over.[11]

THE CRASH OF 1890

The collapse of Barings Bank in London caused turmoil in the financial world known as the "Panic of 1890" and both Westinghouse and Edison found themselves in difficulties as other banks began to call their loans in.

While the Edison empire had undergone a reorganization the previous year and was now in the hands of businessmen, Westinghouse was still pouring money into the hands of inventors. Work on Tesla's induction motors had come to a halt. But Westinghouse and Tesla still had faith in them.

Westinghouse turned up at Tesla's laboratories and asked him to forgo any royalties due under the contract and to give Westinghouse a free hand with the patents so he could resume work.

"Your decision determines the fate of the Westinghouse Company," he said.

"Suppose I should refuse to give up my contract," said Tesla. "What would you do then?"[12]

"In that event you would have to deal with the bankers," said Westinghouse, "for I would no longer have any power in the situation."

"And if I give up the contract you will save your company and retain control so you can proceed with your plans to give my polyphase system to the world?"

"I believe your polyphase system is the greatest discovery in the field of electricity," said Westinghouse. "It was my efforts to give it to the world that brought on the present difficulty, but I intend to continue, no matter what happens, to proceed with my original plans to put the country on an alternating current basis."

Tesla responded with respect:

Mr. Westinghouse, you have been my friend, you believed in me when others had no faith; you were brave enough to go ahead and pay me when others lacked courage; you supported me when even your own engineers lacked vision to see the big things ahead that you and I saw; you have stood by me as a friend. The benefits that will come to civilization from my polyphase system mean more to me than the money involved. Mr. Westinghouse, you will save your company so that you can develop my inventions. Here is your contract and here is my contract – I will tear both of them to pieces and you will no longer have any troubles from my royalties. Is that sufficient? [13]

Or, at least, that was Tesla's account of the interview in later life.

So, after two years, work on Tesla's motors resumed. The young engineer Benjamin Lamme examined the patents and the prototypes, and concluded that Tesla had exhausted all the possibilities of adapting them to run at higher frequencies. He managed to talk his superiors round. Westinghouse would have to go over to 60 cycles per second – the frequency of alternating current used to this day in the US – and the company simply announced that a young engineer named Lamme had discovered the efficiencies of lower frequencies.[14]

WESTINGHOUSE'S FINANCIAL WOES

Westinghouse may not have been so forthright as Tesla made him appear. It is not clear that investing in Tesla's induction motors was the cause of his present financial woes. He had racked up huge legal bills fighting Edison over patent infringements on the incandescent bulb. But Tesla was sanguine. He believed that his work on high-frequency electricity promised new riches.

On May 20, 1891, Thomas Martin, now editor of the *Electrical Engineer*, persuaded Tesla to make another speech to the AIEE. This time it would be called "Experiments with Alternate Currents of Very High Frequency and Their Application to Methods of Artificial Illumination."[15] In it, he said that he had devised an alternating current machine that was capable of giving more than two million reversals a minute.

But for his talk, he confined himself to the use of the electrostatic waves created to produce practical and efficient sources of light, he said. He also took another sideswipe at Edison, remarking that incandescent light bulbs were very inefficient.

"Some better methods must be invented, some more perfect apparatus devised,"[16] he said.

TESLA'S LUMINOUS LECTURE

Tesla then showed his audience a bulb with a single looped filament which spun as it glowed. Then he produced a bulb with no filament at all which glowed as brightly as any incandescent bulb. He held it up, as if he were the Statue of Liberty, and showed that it was not attached by any wire to any source of electricity.

"I suspend a sheet of metal a distance from the ceiling on insulating coils and connect it to one of the terminals of the induction coil, the other terminal being preferably connected to the ground," he explained. "An exhausted tube may then be carried in the hand anywhere between the sheets or placed anywhere, even a certain distance beyond them; it remains always luminous."[17]

The lecture went on for three hours and struck at Edison's first great achievement – the incandescent electric light bulb. But clearly Tesla had more up his sleeve. As he drew his talk to a close, he said:

Among many observations, I have selected only those which I thought most likely to interest you. The field is wide and completely unexplored, and at every step a new truth is gleaned, a novel fact observed. How far the results here borne out are capable of practical applications will be decided in the future. As regards the production of light, some results already reached are encouraging and make me confident in asserting that the practical solution of the problem lies in that direction ... The possibilities for research are so vast that even the most reserved must feel sanguine of the future.[18]

Tesla generated huge sparks and electric flames. Electricity, he showed, would run to earth and it was not as dangerous as Edison made out. As a finale, he ran tens of thousands of volts of AC through his body to light up bulbs and shoot sparks from his fingertips, showing that alternating current was not a killer when handled properly.

TESLA'S ELECTRIC VISION

While Edison was a practical, almost prosaic, inventor, clearly Tesla was a visionary:

> *We are whirling through endless space with an inconceivable speed, all around us everything is spinning, everything is moving, everywhere is energy. There must be some way of availing ourselves of this energy more directly. Then, with the light obtained from the medium, with the power derived from it, with every form of energy obtained without effort, from the store forever inexhaustible, humanity will advance with giant strides. The mere contemplation of these magnificent possibilities expands our minds, strengthens our hopes and fills our hearts with supreme delight.*[19]

The applause was nothing short of rapturous. *Electrical World* declared it "one of the most brilliant and fascinating lectures that it has ever been our fortune to attend."[20] *Harper's Weekly* said: "At one bound he placed himself abreast of such men as Edison, Brush, Elihu Thomson, and Alexander Graham Bell."[21]

Nikola Tesla demonstrating wireless transmission of power and high frequency energy at Columbia College, New York, in 1891. The two metal sheets were connected to his Tesla coil oscillator, which applied a high voltage oscillating at radio frequency. The electric field ionized the gas in the long partially-evacuated Geissler tubes he is holding, causing them to emit light without wires.

LIGHTING WITH NO WIRES

The inventors of early electric lighting knew two ways to produce illumination – either by creating a spark, or arc, between electrodes or by running a current through a wire or fiber, heating it up until it glowed. Arc lamps are very bright and were used in searchlights, floodlights, lighthouses and movie projectors. They were not suitable for domestic use. But using heated filaments also have its drawbacks. Most materials don't behave well when heated near their melting points. They oxidize, unless surrounded by vacuum or inert gas, or break apart through internal stress. Joseph Swan in 1878 and Thomas Edison the following year independently developed the carbon-filament bulb. This was superseded by the more efficient tungsten-filament bulb in 1908.

However, there is another problem with incandescent lamps. In a domestic 60-watt light bulb, for example, no more than a few percent of the energy radiated is visible. Most is lost as heat. But in 1859, Alexandre-Edmond Becquerel (1820 – 91) discovered that certain substances fluoresced when a current was passed through a Geissler tube – that is, a partially evacuated glass cylinder. Tesla developed this into the phosphorescent lamp – phosphorescent substances are slower to emit light than fluorescent ones and continue to glow for some time after the power is switched off. He began by powering conventional filament or arc lamps with high-frequency currents. This caused the diffused gases inside to glow and made certain solid materials give off light. The bulbs remained cold because most of the electrical energy passed through them turned into light, rather than heat. Consequently, they were much more efficient. But although he used these experiments to illustrate his celebrated lectures, he seldom patented them.

Having developed apparatus that produced higher frequencies and voltages than were available to anyone else, by 1890, he was able to light phosphorescent tubes without connecting wires. The energy was transmitted to them at radio frequencies. At higher energies, Tesla's tubes gave off X-rays.

Nikola Tesla demonstrates a glowing, wireless phosphorescent lamp.

EDISON'S SPIRALING EXPENSE

Thomas Edison found himself in financial trouble too. He soon discovered that Villard's reorganization was not an unalloyed blessing. Before it, Edison said he had "an income of $250,000 per year, from which I paid easily my laboratory expenses. This income by the consolidation was reduced to $85,000, which is insufficient to run the laboratory."[22] As a result, he had to spend half his time back in the lighting business.

Barings' difficulties brought about the collapse of Villard's German-funded North American Bank. Behind Edison's back, Villard entered into secret talks with Charles Coffin, who was then running Thomson-Houston. He had long sought a merger between the two companies to put an end to costly patent suits. Meanwhile Coffin was also making unsuccessful overtures to Westinghouse, who had no time for financial trickery.[23]

On July 14, 1891, a Federal court in the Southern District of New York sustained the decision of an earlier Pittsburgh court that the rights to the incandescent light bulb belonged to Edison. Westinghouse appealed, but there was panic among the electrical companies. Half the incandescent lamps in the country were made by other companies who now faced paying Edison $2 million a year in royalties.[24]

HIGH-POWERED FINANCE

In an effort to calm the panic, Martin pointed out in the *Electrical Engineer* that salvation was at hand. In an article headlined "The Edison Lamp Decision," he wrote: "Mr. Tesla gave us a motor without a commutator; and it would be strangely in-keeping if he gave us now a lamp without a filament."[25]

The same issue of the *Electrical Engineer* announced that Westinghouse had saved his company by doing a deal with stockholders and bringing high-powered financiers onto the board.

With Westinghouse now in a position to continue the "Seven Years' Incandescent Light Bulb War," the price of Edison stock dipped from $120 a share to $90. In the Wall Street office of J.P. Morgan, Villard, and Coffin agreed the amalgamation of Edison General Electric and Thomson-Houston.

When Edison's personal secretary, Alfred O. Tate, heard of the deal, he jumped on the Hoboken ferry to New Jersey. Tate later wrote:

> *I have always regretted the abruptness with which I broke the news to Edison, but I am not sure that a milder manner and less precipitate delivery would have cushioned the shock. I never before had seen him change color. His complexion naturally was pale, a clear healthy paleness, but following my announcement it turned as white as his collar.*[26]

Thomson-Houston, though smaller, was much more profitable. Its management were put in charge of the merged company, so essentially Thomson-Houston was buying out Edison General Electric. Consequently, "Edison" was dropped from the name.

EDISON'S ABERRATION

Edison was reportedly "disgusted" by the turn of events, but he put a brave face on it, telling *The New York Times*: "I cannot waste my time over electric-lighting matters, for they are old. I ceased to worry over those things ten years ago, and I have a lot more new material on which to work. Electric lights are too old for me. I simply wish to get as large dividends as possible from such stock as I hold. I am not businessman enough to spend my time at that end of the concern."[27]

But Thomas Martin knew where the blame lay. In an editorial headlined "Mr. Edison's Mistake," he pointed out that Edison's knock back was due to:

> *... the attitude taken, and persistently held, by Mr. Edison towards alternating current distribution. He could see no merit in that*

system. But upon its advent, its possibilities were promptly perceived by others ... Since its introduction for long-distance service, six years ago, it has practically driven the direct system from the field of much central station business. Mr. Edison set his face against it as a flint from the first, and has sought on every possible occasion to discredit it through the weight in the community of his justly great name. But the tide would not turn back at his frown.[28]

VIOLATING EDISON'S NAME

Edison won another judgment in the Seven Years' Light Bulb War in the fall of 1892, but Tate found the great man much diminished. When Tate asked Edison for some technical data, he replied:

Tate, if you want to know anything about electricity go out to the galvanometer room and ask Kennelly. He knows far more about it than I do. In fact, I've come to the conclusion that I never did know anything about it. I'm going to do something now so different and so much bigger than anything I've ever done before, that people will forget that my name ever was connected with anything electrical.

Tate was indeed shocked:

I knew that something had died in Edison's heart and that it had not been replaced by the different and bigger thing to which he had referred. His pride had been wounded. There was no trace of vanity in his character, but he had a deep-seated, enduring pride in his name. And that name had been violated, torn from the title of the great industry created by his genius through years of intensive planning and unremitting toil.[29]

But General Electric now controlled three-quarters of the country's electrical business and there were rumors that it would soon take over Westinghouse.

Portrait of Edison by Abraham Archibald Anderson (1890).

THE TESLA COIL

Tesla had used coils and capacitors when experimenting with rotating magnetic fields for his induction motors. He continually refined the components, particularly the special transformer, or coil, at the heart of the circuit, taking out his first patent for a device to run a new and more efficient lighting system in 1891. The basic circuit connected a power supply to a large capacitor, the coil or inductor, and the electrodes of an adjustable spark-gap. As the capacitor charges up, the voltage lags behind the current. In an inductor, current lags behind the voltage as it has to push against the magnetic field its own passage causes.

If the size of the capacitor and coil are selected so that the voltage peaks in the capacitor just as it reaches a minimum in the coil, current and voltage can be made to chase each other back and forth. This oscillation is initiated by the spark gap. As the voltage in a capacitor builds up, it reaches a level when the air in the gap, which acts as an insulator, breaks down due to ionization and lets current flow.

In a Tesla Coil, the inductor is the primary coil of a transformer. When the circuit sparks, all the energy stored over several microseconds is discharged in a powerful impulse, producing a high voltage in the secondary coil. Once the energy has been discharged, the voltage across the spark gap falls and the air becomes an insulator again, until the voltage in the capacitor builds up again. This whole process can repeat itself many thousands of times in a second.

In a Tesla Coil, the secondary winding is designed to react quickly to a sudden energy spike. The electrical impulses induced in it propagate along the winding as waves. The length of the coil is calculated so that, when the wave crests reach the end and are reflected back, they meet and reinforce the waves behind them, so it appears that the voltage peaks are standing still. If the high-voltage end of the secondary coil is attached to an aerial, it becomes a powerful radio transmitter.

In the early decades of radio, most practicable radio transmitters used Tesla Coils. Tesla himself used larger or smaller versions of his invention to investigate fluorescence, X-rays, radio, wireless power and even the electromagnetic nature of the Earth and its atmosphere.

A Tesla coil discharges flashing purple sparks of electricity during a modern day demonstration.

TESLA'S INTERNATIONAL ROADSHOW

While Edison was licking his wounds in New Jersey, Tesla was establishing his worldwide reputation. He was invited to give two lectures in London in February 1892 – the first in front of the Institution of Civil Engineers, the second before the Royal Institution. Tesla was reluctant to repeat his lecture there, but James Dewar (1842 – 1923), the institution's professor of chemistry, "pushed me into a chair and poured out half a glass of a wonderful brown fluid which sparkled in all sorts of iridescent colors and tasted like nectar. 'Now,' he said, 'you are sitting in Faraday's chair and you are enjoying the whisky he used to drink.' "[30]

Tesla was then to give a lecture on the same stage where Faraday had outlined the fundamental principles of electromagnetic induction in the 1830s. He did not disappoint. He put on another show featuring huge sparks, glowing wires, wireless motors, and colored lights that spelt out the name William Thomson who, that year, became Lord Kelvin. And, to the amazement of his audience of distinguished scientists, Tesla again ran high-volt AC current through his body.

He melted and vaporized tinfoil in a coil. With a new type of lamp, he disintegrated zirconia and diamonds, the hardest known substances at the time. He was so far ahead of his time that he described the ruby laser that would not be built until 1960.

WOWING LONDON AND PARIS

One of his demonstrations involved the vacuum tube which would later be used to amplify weak radio signals. Improvements could be made to the transatlantic telegraph cables so they could carry telephone calls, he said. He also speculated on the future of wireless transmission. The *Electrical Review* said:

> *The lecture given by Mr. Tesla ... will long live in the imagination of every person ... that heard him, opening as it did, to many of them, for the first time, apparently limitless possibilities in the applications and control of electricity. Seldom has there been such a gathering of all the foremost electrical authorities of the day, on the tiptoe of expectation.*[31]

Tesla made a rod for his own back when he "tantalizingly informed listeners that he had showed them but one-third of what he was prepared to do."[32] As a result, the audience remained in their seats and he was forced to deliver a supplementary lecture and an encore. Then he presented Lord Kelvin with one of his early experimental Tesla coils which would become a key component in the development of wireless transmission.

Tesla moved on to Paris where he wowed French academicians at the *Société International de Physique* and the *Société International des Electriciens.*

"The French papers this week are full of Mr. Tesla and his brilliant experiments," reported the *Electrical Review.* "No man in our age has achieved such a universal scientific reputation in a single stride as this gifted young electrical engineer."[33]

Then he headed home to Gospic where he found his mother gravely ill. She died soon after. He later wrote: "The mother's loss grips one's head more powerfully than any other sad experience in life."

TESLA IS "THE NEW EDISON"

While in Europe, Tesla visited the Ganz works in Budapest to see a 1,000-volt alternator they were building. He went to Berlin to visit Helmholtz, who developed the mathematics of electrodynamics. Then he went on to Bonn to see Heinrich Hertz, the first man to transmit and receive radio waves. Hertz had conducted his experiments with a simple sparking apparatus that could transmit radio waves across his lab.

While Hertz was a theoretical physicist who simply wanted to investigate the theories of James Clerk Maxwell, Tesla was an electrical engineer who wanted to put them to a practical use. He had already duplicated Hertz's experiments, developing from this work the Tesla coil which was capable of transmitting wirelessly over long distances.

Hertz inferred the existence of a mysterious substance called ether that permeated the whole universe. The reasoning was that, if light and other electromagnetic phenomena are waves, they must have something to propagate through. Tesla maintained that such a substance could not exist as it would have to be both inconceivably tenuous and extremely rigid. They fell out over the matter. In fact, the existence of ether had already been disproved by A.A. Michelson (1852 – 1931) in Germany in 1881 and, again, in collaboration with Edward Morley (1838 – 1923) in the US in 1887 in their famous Michelson-Morley experiment.

When Tesla arrived back in New York after his triumphant trip to Europe, he was greeted with a photograph of Edison signed: "To Tesla from Edison."[34] Then Westinghouse dropped by with the news that they had won the contract to provide the power for the forthcoming 1893 World's Fair in Chicago.[35]

TESLA VERSUS HERTZ

While Tesla was already being hailed as the "new Edison," he was determined to take on Hertz as well. He noted that Hertz had conducted his experiments on radio transmission with a battery and a simple circuit interrupter, like a Morse key, connected to an induction coil – a small transformer – to produce a high-voltage spark. This could be detected using a copper loop with a spark gap. Tesla quickly realized that, instead of a battery with a circuit interrupter, it would be better to use an AC current. While a circuit interrupter would only give a frequency of, at best, a few hundred cycles per second, an alternator could give ten- or twenty-thousand cycles per second. As a mechanical device, when an alternator reached that speed it would begin to fly apart. However, higher frequencies could be generated electrically.

Tesla was already ahead of the field in this area. He had used induction coils and

HEINRICH HERTZ
(1857 – 94)

After studying under Helmholtz in Berlin, Hertz began his investigation of the theories of James Clerk Maxwell in 1883. At Karlsruhe Polytechnic, he developed primitive equipment to generate electromagnetic waves and measured their wavelengths and velocity. Showing that they could be reflected and refracted like light and radiant heat, he demonstrated that light and heat were also electromagnetic waves. He was just thirty-six when he died. The frequency of electromagnetic waves are measured in hertz – one hertz being equal to one cycle per second.

capacitors – devices such as Leyden jars to store static electricity – to give split-phase AC currents to run his induction motors. Putting a connecting capacitor across the terminals of a coil produced a circuit that resonated, giving a spike in output. Tesla called this an oscillating transformer, though others began calling it the Tesla Coil. A coil coupled to a capacitor resonating at a specific frequency is the basis of all wireless transmission.

Similar tuned circuits could be used as detectors, along with vacuum tubes. Earthing one terminal of the oscillating transformer to the city's water main, he moved around New York detecting the electromagnetic waves generated at various frequencies. However, his idea was not to use radio waves to transmit information as we do now. His goal was to transmit electric power *wirelessly*.[36]

THE SKIN EFFECT

In his early experiments, Tesla had discovered the "skin effect." He had accidentally touched a high-voltage terminal, but remained, to his surprise, unhurt. This was because, at high frequencies, the magnetic field generated pushes the current to the outside of a conductor. So instead of running through his body, it travels across the surface, leaving the internal structure undamaged. In his public demonstrations, he touched one terminal of a high-frequency apparatus generating tens of thousands of volts and illuminated a bulb or tube held in the other hand. Again, in a swipe at Edison, he also showed that an alternating current, if at a sufficiently high frequency, was safer than a direct current.

Nikola Tesla holding in his hand balls of flame.

FATHER OF WIRELESS TRANSMISSION

Tesla's lectures attracted so much interest that he took them on the road, drawing huge audiences with his dazzling demonstrations. In Philadelphia, in 1893, he outlined a method of transmitting moving pictures – which we know as television. He had discovered that the secret of wireless transmission was resonance. Electrical impulses jump from a sending device to a receiver through thin air if they are tuned to the same frequency. As part of the show, he unveiled a diagram showing aerials, transmitters, receivers and earth connections, all the elements of a modern broadcast system.

It was not just the theory that thrilled, but his practical demonstrations. On one side of the stage would stand his transmitter. This was a high-voltage transformer connected to a bank of Leyden jars and a coil – giving a tuned circuit – attached to a sparking gap and a length of wire hanging from the ceiling, acting as an aerial. On the other side there was an identical length of wire. The receiving aerial was connected to an identical coil and bank of Leyden jars, giving a circuit tuned to the same frequency. But instead of using a sparking gap as a detector, Tesla used a Geissler, or discharge, tube which glowed when electricity passed through it, like a primitive neon light.

The demonstration was accompanied by strange sounds. When electricity was fed to the transformer, the magnetic field created made the core strain, producing odd groaning sounds. Sparks cracked across the sparking gap and corona sizzled around the edges of the foil on the Leyden jars as the radio waves traveled noisily from one antenna to the other. Then, magically, the Geissler tube lit up.

Although Tesla was convinced of the possibilities of his wireless system, he was advised to play them down. Sponsors feared such fanciful ideas may deter conservative businessmen who might otherwise be interested in his lighting systems or his motors. Nevertheless, fellow researchers were already calling him the "Father of the Wireless."[37] While serious physicists had investigated the phenomenon of wireless transmission before him, Tesla had pioneered the use of the tuned circuit, the aerial and the ground connection. What's more he was giving these demonstrations of wireless in 1893, a year before Guglielmo Marconi even began his experiments.

Three years later, Tesla received a request from Sir William Preece, an engineer with the Imperial Post Office in London, for two wireless sets for tests. But Marconi had moved from Italy to London by then. He told Preece not to bother with the Tesla system. He had tried it and it had not worked. Unperturbed, Tesla filed a patent for wireless transmission in September 1897.[38]

WORKING WITH THE DEVIL

St. Louis, Missouri, went wild for Tesla. Four thousand copies of an electrical journal, which normally had a small circulation, were sold because it carried an article about him. When Tesla arrived in town, there was a parade of eighty electrical utility wagons. The four-thousand-seat Grand Music Entertainment Hall was filled to overcapacity as several thousand more squeezed in and scalpers were selling more tickets in the streets outside. The audience were not disappointed when Tesla passed 200,000 volts through his body. The fireworks were accompanied by a public lecture:

I now set the coil to work and approach the free terminal with a metallic object held in my hand, this simply to avoid burns. As I approach the metallic object to a distance of eight or ten inches, a torrent of furious sparks breaks forth from the end of the secondary wire, which passes through the rubber column. The sparks cease when the metal in my hand touches the wire. My

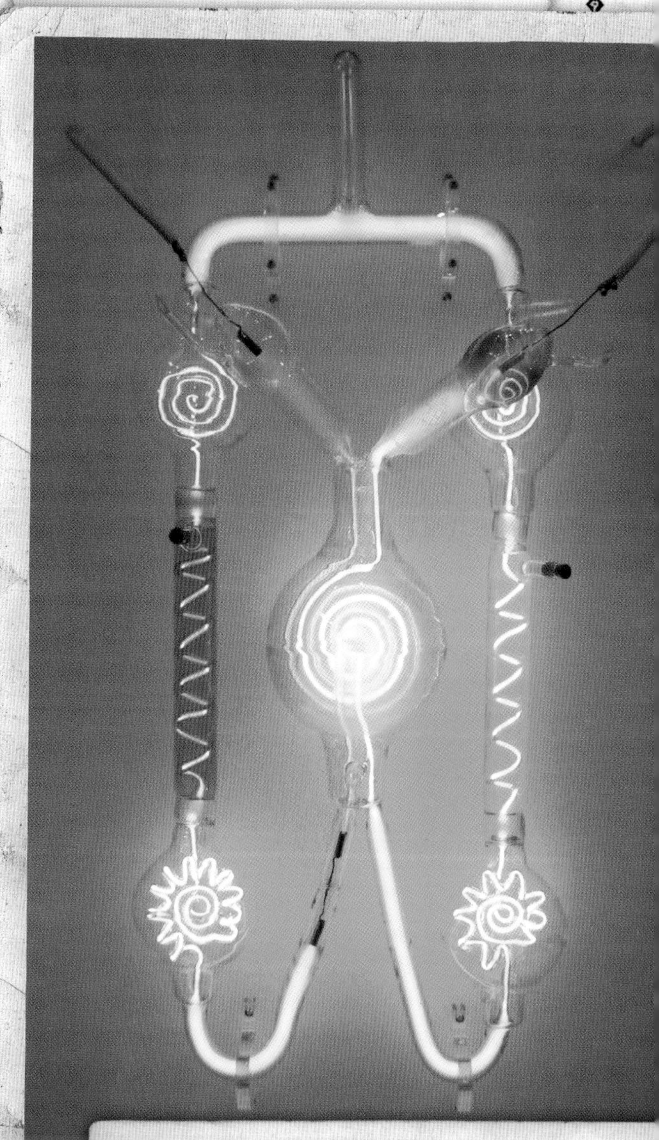

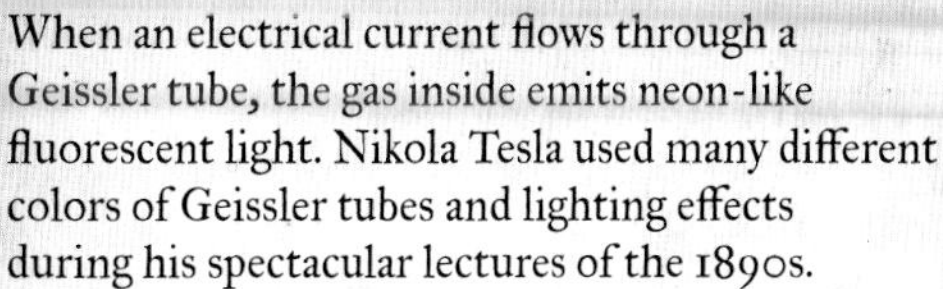

When an electrical current flows through a Geissler tube, the gas inside emits neon-like fluorescent light. Nikola Tesla used many different colors of Geissler tubes and lighting effects during his spectacular lectures of the 1890s.

GUGLIELMO MARCONI
(1874 – 1937)

In 1894, Marconi began experimenting with an induction coil, a Morse key and a sparking gap, along with a simple detector, at his father's estate near Bologna, Italy. Devising a simple aerial and an improved coherer – a primitive detector – he increased the range to 1.5 miles (2.4 kilometers).

He moved to London, England, where he filed his first patent in June 1896. Using balloons and kites, he increased the range still further. In 1899, signals were sent across the English Channel and the America's Cup used Marconi's equipment for ship-to-shore communication. The following year, Marconi took out patent No. 7777, which enabled several stations to operate on different frequencies. This was overturned by the US Supreme Court in 1943, when it was shown that Tesla and others had already developed radio-tuning circuits.

In December 1901, Marconi transmitted a signal across the Atlantic from Cornwall, England, to Newfoundland, Canada. The curvature of the Earth had proved no obstacle because radio waves reflected off ionized layers in the upper reaches of the atmosphere. Marconi continued to improve the range and efficiency of wireless devices and set up companies to exploit his discoveries. In 1909, he was awarded the Nobel Prize for physics and in 1932, the Marconi company won the contract to establish short-wave communication between England and the countries of the British Empire.

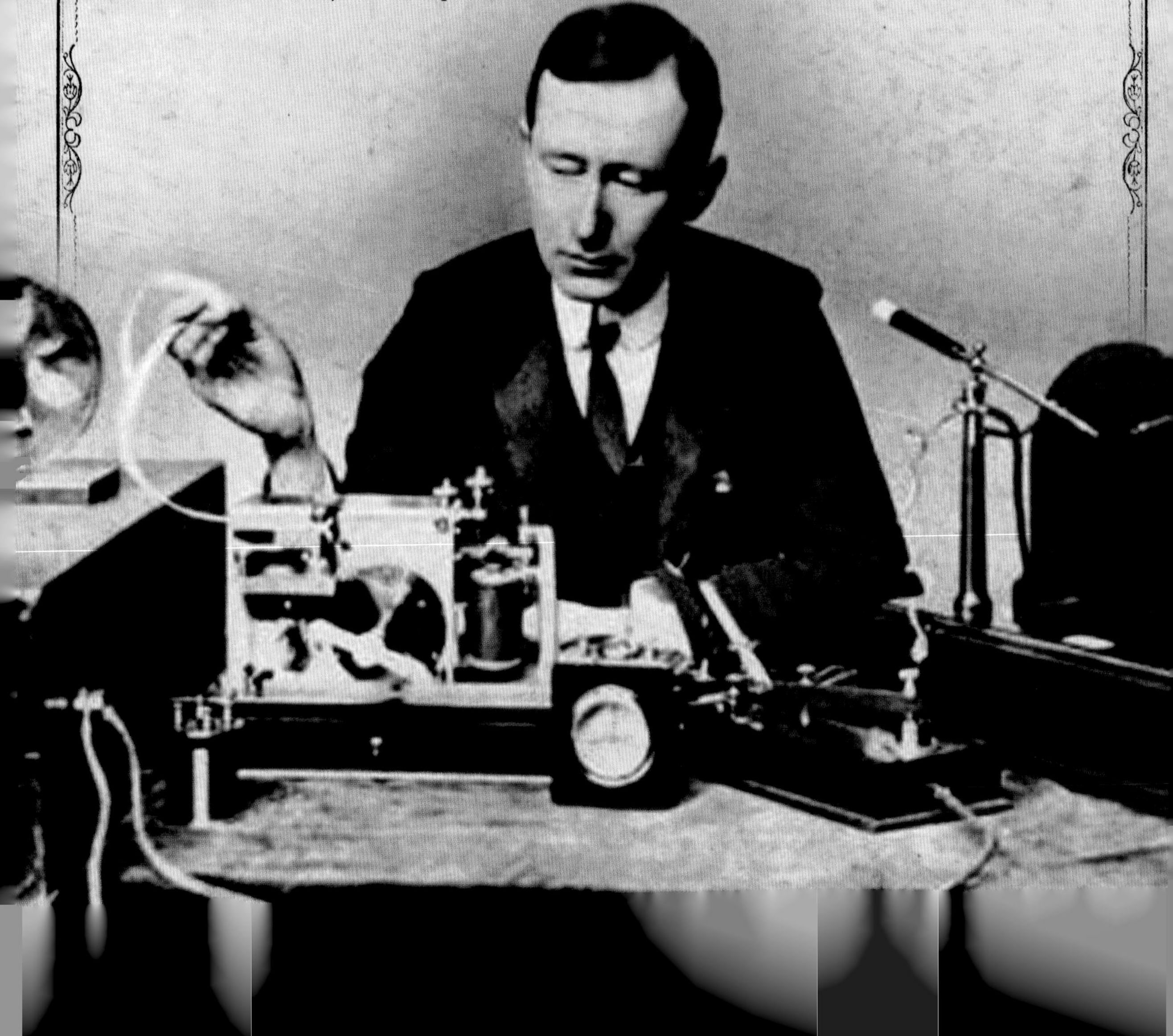

arm is now traversed by a powerful electric current, vibrating at about the rate of one million times a second. All around me the electrostatic force makes itself felt, and the air molecules and particles of dust flying about are acted upon and are hammering violently against my body.

So great is this agitation of the particles, that when the lights are turned out, you may see streams of feeble light appear on some parts of my body. When such a streamer breaks out on any part of the body, it produces a sensation like the pricking of a needle. Were the potentials sufficiently high and the frequency of the vibration rather low, the skin would probably be ruptured under the tremendous strain, and the blood would rush out with great force in the form of fine spray or jet so thin as to be invisible, just as oil will when placed on the positive terminal of a Holtz machine [electrostatic generator]. The breaking through of the skin though it may seem impossible at first, would perhaps occur, by reason of the tissues under the skin being incomparably better at conducting. This, at least, appears plausible, judging from some observations.

I can make these streams of light visible to all, by touching with the metallic object one of the terminals as before, and approaching my free hand to the brass sphere, which is connected to the second terminal of the coil. As the hand is approached, the air between it and the sphere, or in the immediate neighborhood, is more violently agitated, and you see streams of light now break forth from my fingertips and from the whole hand. Were I to approach the hand closer, powerful sparks would jump from the brass sphere to my hand, which might be injurious. The streamers offer no particular inconvenience, except that in the ends of the fingertips a burning sensation is felt ...

The streams of light which you have observed issuing from my hand are due to a potential of about 200,000 volts, alternating in rather irregular intervals, sometimes like a million times a second. A vibration of the same amplitude, but four times as fast, to maintain which over three million volts would be required, would be more than sufficient to envelop my body in a complete sheet of flame. But this flame would not burn me up; quite contrarily, the probability is, that I would not be injured in the least. Yet a hundredth part of that energy, otherwise directed; would be amply sufficient to kill a person ...[39]

Waving shaped glass tubes in the powerful electromagnetic field created by his oscillating transformer, Tesla created stunning effects like the "spokes of a wheel of glowing moonbeams,"[40] the *Electrical Engineer* said.

His lights needed no connections to make them glow. Towards the end of the performance, Tesla held up one of his phosphorescent tubes, the precursor of fluorescent lights. He then announced that he would illuminate it simply by touching the terminal of his oscillating transformer with his other hand. When he did, the lamp lit up.

"There was a stampede in the two upper galleries and they all rushed out," said Tesla. "They thought it was some part of the devil's work."[41]

PAY BACK TIME FOR TESLA

After his European tour, Tesla acquired US citizenship. He was now a "full-fledge American" and decided to see whether Edison could take an American joke. He devised a demonstration that pitted the carbon-filament incandescent light Edison had invented against an identical bulb that was empty. When he applied a current at a frequency of around one million cycles per second to the empty bulb it glowed – more brightly than the Edison bulb which was being run on direct current. What's more, while the incandescent bulb was hot, the empty bulb stayed cool to the touch. Edison was far from pleased and, again, Tesla had outsmarted his former mentor in the popular press.

Nikola Tesla zipping wireless light in his New York lab in 1898.

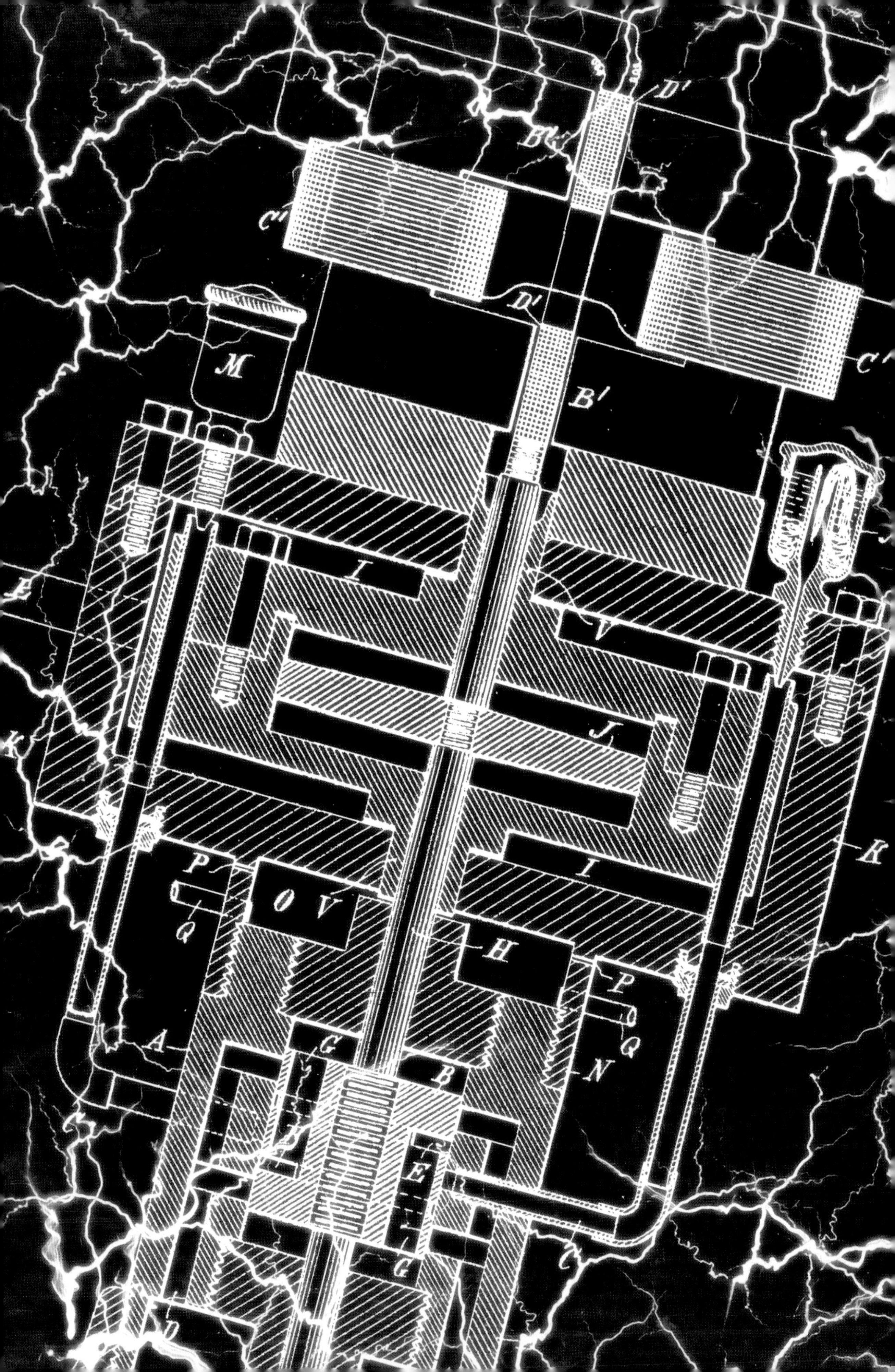

D'
B'
C'
D'
C'
M
B'
I
V
J
K
K
P
O V
I
Q
H
P
A
G
B
Q
N
E
G
C
D

PART FOUR
WIRELESS TRANSMISSION OF POWER
E'

CHAPTER 10

CHICAGO WORLD'S FAIR 1893

The second round of the War of the Currents was to take place at the World's Columbian Exposition, aka the Chicago World's Fair, in May 1893. Westinghouse had won the contract to light it by putting in a bid much lower than that of General Electric, which now owned Edison's patents. The site of the Fair was to be illuminated with two hundred thousand bulbs spread out over more than a square miles. This provided the ideal opportunity to demonstrate how Tesla's AC system could be used to light an entire city.

Initially, Westinghouse was not in the bidding. General Electric put in a bid of $38.50 per arc light for six thousand lights, which were still the best way to illuminate large public spaces. But, before the merger, Edison had charged just $11 an arc light to illuminate the construction site of the Fair for night work.

The Fair's directors were not amused and organized a number of smaller companies to provide arc lights at $20 a piece. But they still needed electricity. Charles Coffin at General Electric offered to provide dynamos at $15.78 a horsepower. Again the Fair directors found a smaller company that would provide electricity at $2.50 a horsepower.

Next came the contract for the 92,000 incandescent lights to illuminate the site. GE

The Chicago World's Fair lit by electricity in May, 1893.

put in a bid of $18.50 a lamp, or $1,702,000. But local businessman Charles F. Locksteadt put in a bid of $6.80 a lamp, or $625,600. But where was he going to get the lamps? He approached Westinghouse.[1]

UNDERCUTTING GENERAL ELECTRIC

Until this point, Westinghouse had been concentrating all his energies on saving his company. Now he realized that he had in front of him, the greatest possible showcase for AC. Westinghouse traveled to Chicago in his private railroad car, arriving to the fanfares of the press.

GE dropped its bid to $6 a lamp, but Westinghouse then promised to underbid them. When the day for the awarding of the final contracts arrived and the bids were unsealed, Westinghouse undercut GE by $80,000.

"There is not much money in the work at the figures I have made," Westinghouse told the press, "but the advertisement will be a valuable one and I want it."[2]

General Electric then pointed out that the matter of the patents was still before the courts. So Westinghouse reluctantly came up with a $1-million bond guaranteeing the contract.

TURNING ENGINEERING ON ITS HEAD

At the price he was being paid, Westinghouse had to devise a more economical system. Back in Pittsburgh, he told his top draftsman E.S. McClelland to design an engine of 1,200 brake horsepower to run at 200 revolutions per minute, when engines that size usually ran at 75 rpm.

"We were building 250 horse power engines," said McClelland. "A 1,200 brake horsepower engine to operate at 200 revolutions per minutes seemed to me to be entirely out of all reason."[3]

McClelland's boss in the drafting department said it was impossible, but McClelland and a colleague stayed up half the night working on it. Then early next day, the drafting department reassembled to look at their drawing. To get a better view, they turned the drawing board on its side.

"This setting of the board on end, strange as it may seem, gave us the solution to the problem," said McClelland. "As a vertical engine there was space to spare. When this solution flashed upon our minds the leading engine draftsman seemed to be electrified and became wildly enthusiastic."

When Westinghouse strode in, he asked: "How soon may I have four of them?"[4]

Straightaway, work began on steam engines and electric generators of an entirely new and untested design.

PERFECTING TESLA'S POLYPHASE

Westinghouse then set about redesigning the light bulb so that it would not infringe any of Edison's patents. Instead of being a sealed unit, his design carried a low-resistance filament on a stopper that fitted into a glass globe filled with nitrogen like a cork into a bottle. Not only could the stopper be removed and the filament replaced, his bulb could be made almost entirely by machine and would be cheaper than the Edison lamp.

Additionally, Westinghouse could still use the Edison-style lamps as the disputed patents were now on their way to the US Supreme Court. Besides, Edison's patent ran out in 1894. Meanwhile Westinghouse launched another assault in the courts, insisting that GE be investigated under Sherman Anti-Trust Act of 1890.

Alerted to GE's intention of getting a restraining order against Westinghouse's "stopper" bulbs, Westinghouse got a ruling that it was no infringement on Edison's patents. He manufactured a quarter of a million of them.[5]

Charles Scott suggested that they stick to tried-and-tested single-phase AC to illuminate them, but Tesla had told Westinghouse that his motors needed two-phase AC, so Westinghouse engineers set out to create viable two-phase generators. Scott recalled:

> *Commercial circuits were single-phase at a frequency of 133 cycles. Strenuous efforts to adapt the Tesla motor to this circuit were in vain. The little motor insisted on getting what it wanted, and the mountain came to Mahomet. Lower-frequency polyphase generators inflicted obsolescence on their predecessors in a thousand central stations – such was the potency of the Tesla motor.*[6]

Sixty cycles per second has been used for lighting ever since and thirty for motors. Proved right, Tesla borrowed equipment from Pittsburgh to work on fresh improvements on his motors, so they would be ready to be exhibited at the Fair.

The Westinghouse (Sawyer-Man) "stopper lamp."

THE STAR OF THE SHOW

In less than six months, they designed and built bigger generators than had ever been built before. Westinghouse built the biggest AC central station in America. It could power 160,000 lamps, along with many motors. Until then the biggest had powered ten thousand lamps and no motors.

Using AC at high-voltage, they could distribute this throughout the Fair on thin wires, saving hundreds of thousands of dollars worth of copper. The Fair site would be a blaze of light, consuming three times the amount of electricity then being utilized by the whole of the city of Chicago. This showcase was now to promote Tesla's motors and his polyphase system, so Tesla went to Pittsburgh, he said, "to bring the motor to high perfection."[7]

According to Benjamin Lamme, who did much of the work: "It was at Mr. Westinghouse's suggestion that the machines for the lighting plant at Chicago were each made with two single phase alternators, side-by-side, with their armature windings staggered 90 degrees."[8]

As well as creating Tesla's two-phase current, each unit would power thirty thousand lamps, and if one unit failed, the other would kick in. The huge engines which reached Chicago only weeks before the Fair opened became the largest exhibit in the exposition, indeed the largest display of operating machinery that had ever been shown.

A DAZZLING SPECTACULAR

The Columbian Exposition covered almost 700 acres (283 hectares) and attracted some twenty-eight million visitors from all over the world. The center-piece was a Ferris wheel standing 264 feet (80 meters) high that could carry over two thousand people. It revolved on the largest one-piece axle ever forged.

But it was Westinghouse's illuminations that took the breath away. Illinois' former governor, Will E. Cameron, described his amazement:

Inadequate words have been found to convey a realizing idea of the beauty and grandeur of the spectacle which the Exposition offers by day, they are infinitely less capable of affording the slightest conception of the dazzling spectacle which greets the eye of the visitor at night ... Indescribable by language are the electric fountains. One of them, called "The Great Geyser," rises to a height of 150 feet [45 meters], above a band of "Little Geysers" ... so bewildering no eye can find the loveliest, their vagaries of motion so entrancing no heart can keep its steady beating.[9]

THE ELECTRICITY PAVILION

At the Chicago World's Fair, the Electricity Pavilion rose to 169 feet (52 meters) and covered the ground space of two soccer pitches. AEG exhibited the equipment they had used to transmit AC current the world-record distance of 109 miles (175 kilometers) from Lauffen to Frankfurt in Germany. GE also exhibited their new AC system. Both were infringing Tesla's patents, but Westinghouse lodged no complaint as they clearly showed the superiority of AC. Instead, they erected a 45-foot (14 meter) high monument to the "Westinghouse Electric & Manufacturing Co. Tesla Polyphase System."[10]

GE also erected an 82-foot (25-meter) Tower of Light in the center of the Electricity Pavilion with eighteen thousand bulbs around the base. It was topped by a huge Edison light bulb. This was supposed to celebrate Edison's victory in the Seven Years Incandescent Light Bulb War – also on display were seven volumes containing seven-thousand pages of testimony in the "Filament Case."[11]

GE was still eager to associate itself with Edison, who exhibited his phonograph, the multiplex telegraph and his Kinetoscope, which produced moving pictures for an individual viewer. Some 2,500 different kinds of Edison lamps were on display. However, Coffin insisted that the company also exhibit some of its own AC equipment.

THE AGE OF ELECTRICITY

The undoubted theme of the Exposition was electricity. The Fair's chief electrician said:

The Columbian Exposition is a magnificent triumph of the Age of Electricity ... all the exhibits in all the buildings are operated by electrical transmission. The Intramural Elevated Railway, the launches that ply the Lagoons, the Sliding Railway on the thousand-foot pier, the great Ferris Wheel, the machinery of the Libby Glass Company on the Midway, are all operated by electrically transmitted energy ... everything pulsates with quickening influence of the subtle and vivifying current.[12]

There was even an electric kitchen.

Just four years earlier, the Paris Exposition of 1889 had been illuminated with 1,150 arc lights and ten thousand incandescent lamps. Chicago had ten times that number. Nor were there wires strung overhead. They were carried in underground tunnels tall enough for a man to stand up in and accessed by 1,560 manholes.

On the Westinghouse stand, Tesla exhibited AC motors and generators, and had the names of famous electrical pioneers – Benjamin Franklin, Helmholtz, Faraday, Maxwell, and Henry – spelt out in phosphorescent tubes, along with that of American physicist John Henry (1797 – 1878), who gave his name to the unit of inductance, and Serbian poet Jovan Jovanovíc Zmaj (1833 – 1904). Neon signs saying "Westinghouse" and "Welcome Electricians" were lit by discharges of artificial lightning that made a deafening sound. Among the flashing sparks and the tubes lit wirelessly was a large Egg of Columbus spinning furiously.[13] This one was the size of an ostrich egg and was surrounded by smaller copper "planets" which demonstrated his

WESTINGHOUSE ELECTRIC &
COLUMBUS

The Electricity Pavilion of the Columbian Exposition, 1893.

THE KINETOSCOPE

On October 17, 1888, Edison filed a caveat describing his idea. "I am experimenting upon an instrument which does for the eye what the phonograph does for the ear, which is the recording and reproduction of things in motion, and in such a form as to be both cheap, practical and convenient."[14]

He called it the Kinetoscope, using the Greek words "kineto" meaning "movement" and "scopos" meaning "to watch." However, the practical work of turning the idea into reality was largely done by William Dickson, an assistant at West Orange. A working model was ready by 1891.

In it, a strip film passed rapidly between an electric light bulb and a lens, while the viewer peered through a peephole. Behind the eyehole was a wheel with a slot in it. As the wheel spun, this acted as a shutter. When the slot and the frames of the film synchronized, it reproduced the semblance of life-like movement.

At the Chicago World's Fair, Edison showed a film of British prime minister William Gladstone making a speech in the House of Commons. Edison eventually developed this into the Projecting Kinetoscope.

Cutaway image of the Edison Kinetoscope, designed by William Dickson.

theory of planetary motion. The *Electrical Experimenter* said:

In this experiment, one large, and several small brass balls were usually employed. When the field was energized all the balls would be set spinning, the large one remaining in the center while the small ones revolved around it, like moons about a planet, gradually receding until they reached the outer guard and raced along the same.

But the demonstration which most impressed the audiences was the simultaneous operation of numerous balls, pivoted discs and other devices placed in all sorts of positions and at considerable distances from the rotating field. When the currents were turned on and the whole animated with motion, it presented an unforgettable spectacle. Mr. Tesla had many vacuum bulbs in which small, light metal discs were pivotally arranged on jewels and these would spin anywhere in the hall when the iron ring was energized.[15]

Other stands exhibited electric body invigorators, charged belts offering a better sex life, and electric hairbrushes. It was then thought that electricity could cure all ills. Elihu Thomson exhibited a high-frequency coil that could produce a spark five feet (1.5 meters) long. Alexander Graham Bell launched a telephone that transmitted sound on a beam of light, while Elisha Gray (1835 – 1901) unveiled a prototype fax machine

THE TOWER OF LIGHT

The General Electric exhibit opened with John Philip Sousa's band playing his "Picadors March" and a spectacular light show, featuring the Tower of Light, which also became known as "Edison's Column." According to the *Chicago Daily Tribune*:

> *Electricity danced up and down and all about its circumference in time with the rhythm of the music. First, slender lines of purple fire ran straight up and down its great height. Between them, at the next measure, came waves of crimson flame, and then, cutting up the spaces between in geometric figures and circling the column from top to bottom, clear dazzling whites.*[16]

The crowd gathered to witness it chanted: "E-di-son, E-di-son, E-di-son."

Edison's Electric Tower of Light, Chicago World's Fair in 1893.

TESLA'S POLYPHASE SYSTEM

In its report on the Fair, the *Electrical Engineer* spotted what it thought would be the way of the future:

> *In a space marked by the "Golden Pavilion" toward the north end of the building is a complete power transmission plant using the Tesla polyphase system ... within this one space are shown a generating station, a high tension transmission circuit about 30 feet long and a receiving and distributing station. The first contains a 500 h.p. two-phase alternating current generator, a 5 h.p. direct current exciter, a marble switchboard and the necessary step-up transformers. In practice the generator and exciter would naturally be driven by water power ... Both generator and wheel are driven by a 500 h.p. Tesla polyphase motor with a rotating field, and the exciter by a 5 h.p. motor of the same type, both operated by current from the large two-phase alternators in Machinery Hall.*[17]

Tesla's Polyphase Alternating Current 500 horsepower generator.

THE WONDERS OF TESLA

In *Electricity at the Columbian Exposition*, J.R. Barrett described the Exhibit of Tesla Apparatus for High Frequency Experiments:

In one of the Westinghouse spaces was erected a room about 25 feet square, in which were daily shown experiments illustrating the remarkable results obtained by Mr. Tesla in the use of high-frequency, high-potential currents. The current was taken from the lighting circuit connected with the plant in Machinery Hall, and transformed up to an exceedingly high voltage by transformers situated back of the room. The rate of alternation was increased from 7,200 per minute to 400,000 per second, by the use of condensers and an arc-breaking device.

Within the room was suspended two hard-rubber plates covered with tin foil. These were about fifteen feet apart, and served as terminals of the wires leading from the transformers. When the current was turned on, the vacuum bulbs or tubes, which had no wires connected to them, but lay on a table between the suspended plates, or which might be held in the hand in almost any part of the room, were made luminous. These were the same experiments and the same apparatus shown by Mr. Tesla in London about two years ago, where they produced so much wonder and astonishment.

Over the entrance to this room there was a most interesting effect. A glass sign, bearing in its center the name "Westinghouse," was supplied with a current of high-potential, one terminal of the transformer being connected to the letters by means of invisible wires, while the other terminal was connected to tin foil on the back of the glass plate.

When the current was turned on, the electricity was discharged from the letters across the surface of the plate, and over its edge to the tin foil on the back, producing the effect of a modified lightning discharge, and accompanied by a similar deafening noise. This was probably one of the most novel attractions in a sensational way seen in the building, as the noise could be heard anywhere within the Electricity building and the flash of the miniature lightning was very brilliant and startling.[18]

called the teleautography – for a few cents, you could have your signature reproduced electronically at a distance.

THE WIZARD OF PHYSICS

Tesla visited the World's Fair in August to put on a week of demonstrations and to attend the International Electrical Congress being held there. Its honorary chairman was Hermann von Helmholtz, and Tesla showed him his exhibit on the Westinghouse stand personally. A thousand electrical engineers attended, including most of the leaders in the field. Ten dollars were offered for seats to see Tesla, who was introduced as the "Wizard of Physics."[19] However, entrance was limited to those who could produce the appropriate credentials.

According to the *Chicago Tribune*: "The great majority of them came with the expectation of seeing Tesla pass a current of 250,000 volts through his body and perform the marvelous feat with lamps lighted through his body that set Paris wild."[20]

Tesla demonstrated mechanical oscillators and steam generators that were so small it was said they could fit in the crown of a hat. He produced motors that could run so precisely they could be used as electric clocks and a continuous-wave radio transmitter, the implications of which were lost on most of his distinguished audience.

Nevertheless what the Columbian Exposition proved to its twenty-eight million visitors was that AC – the executioner's current – was safe. From then on, over eighty percent of all electrical devices bought in the US worked on alternating current.

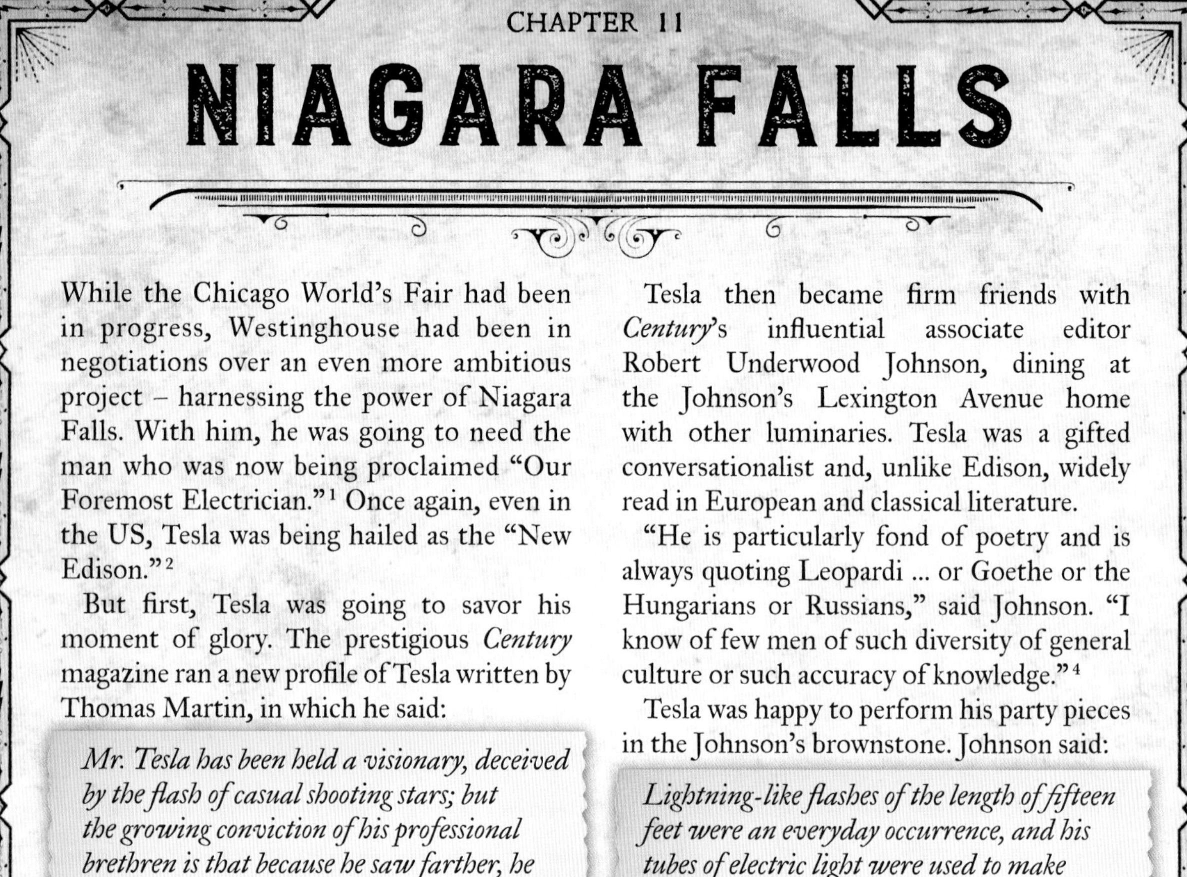

CHAPTER 11

NIAGARA FALLS

While the Chicago World's Fair had been in progress, Westinghouse had been in negotiations over an even more ambitious project – harnessing the power of Niagara Falls. With him, he was going to need the man who was now being proclaimed "Our Foremost Electrician."[1] Once again, even in the US, Tesla was being hailed as the "New Edison."[2]

But first, Tesla was going to savor his moment of glory. The prestigious *Century* magazine ran a new profile of Tesla written by Thomas Martin, in which he said:

> *Mr. Tesla has been held a visionary, deceived by the flash of casual shooting stars; but the growing conviction of his professional brethren is that because he saw farther, he saw first the low lights flickering on tangible new continents of science.*[3]

Tesla then became firm friends with *Century*'s influential associate editor Robert Underwood Johnson, dining at the Johnson's Lexington Avenue home with other luminaries. Tesla was a gifted conversationalist and, unlike Edison, widely read in European and classical literature.

"He is particularly fond of poetry and is always quoting Leopardi ... or Goethe or the Hungarians or Russians," said Johnson. "I know of few men of such diversity of general culture or such accuracy of knowledge."[4]

Tesla was happy to perform his party pieces in the Johnson's brownstone. Johnson said:

> *Lightning-like flashes of the length of fifteen feet were an everyday occurrence, and his tubes of electric light were used to make photographs of friends as souvenirs of their visits. He was the first person to make use of phosphorescent light for photographic purposes – not a small item of invention in itself.*

Mark Twain with an experimental vacuum lamp in Tesla's New York lab, with Tesla in the background.

Railroad-magnate William "Willie" K. Vanderbilt (1849 – 1920) would lend Tesla the Vanderbilt box at the Metropolitan Opera House. Mark Twain visited Tesla's laboratory and Tesla visited Rudyard Kipling, who was then living in Vermont. Architect Stanford White nominated him for the Player's Club in Gramercy Park.

With Tesla's help, Martin published *The Inventions, Researches and Writings of Nikola Tesla* in 1894. Tesla got honorary doctorates from Columbia and Yale. His voice was recorded on a phonograph, an honor already bestowed on the Australian opera singer Nellie Melba (1861 – 1931) and Sarah Bernhardt.

GREATER EVEN THAN EDISON

Tesla sat for a sculptor and journalists flocked to interview him. Joseph Pulitzer (1847 – 1911), who later established the Pulitzer prizes but was then publisher of the *New York World*, sent a young reporter named Arthur Brisbane (1864 – 1936) to write a profile of Tesla. It was headlined: "OUR FOREMOST ELECTRICIAN." The subhead ran: "Greater Even Than Edison." Tesla was then hailed as the master of the "electricity of the future" and the article was accompanied by a full-length drawing of Tesla radiating what the caption called "the effulgent glory of myriad tongues of electric flame after he has saturated himself with electricity."[5]

The interview took place at Delmonico's. Brisbane noted the famous restaurateur lowered his voice at the mention of Tesla's name. According to Brisbane, Charles Delmonico said:

> *That Tesla can do anything. We managed to make him play pool one night. He had never played, but he had watched us for a little while. He was very indignant when he found that we meant to give him fifteen points. But it didn't matter much, for he beat us all even and got all the money. There are just a few of us who play for 25 cents, so it wasn't the money we cared about, but the way he studied out pool in his head, and then beat us, after we had practiced for years, surprised us.*[6]

When asked what it was like to subject himself to such huge voltages, Tesla said: "I admit that I was somewhat alarmed, when I began these experiments, but after I understood the principles, I could proceed in an unalarmed manner."[7]

BRISBANE ON TESLA

"When Mr. Tesla talks about the electrical problems upon which he is really working he becomes a most fascinating person. Not a single word that he says can be understood. He divides time up into billionths of seconds, and supplies power enough from nothing apparently to do all the work in the United States. He believes that electricity will solve the labor problem. That is something for Mr. [Eugene] Debs [the imprisoned labor leader] to ponder while he languishes in his dungeon. It is certain, according to Mr. Tesla's theories, that the hard work of the future will be the pressing of electric buttons."

Arthur Brisbane

New York World

July 22, 1894

A Westinghouse-Tesla Niagara generator under construction in Pittsburgh.

It was to get across the principles that he continued to use the demonstration in his lectures.

My idea of letting this current go through me was to demonstrate conclusively the folly of popular impressions concerning the alternating current. The experiment had no value for scientific men. A great deal of nonsense is talked and believed about 'volts' etc ... You see, voltage has nothing to do with the size and power of the current.[8]

He later explained the spectacle presented when he was connected to an AC voltage of two-and-a-half million volts. It was, Tesla said:

... a sight marvelous and unforgettable. One sees the experimenter standing on a big sheet of fierce, blinding flame, his whole body enveloped in a mass of phosphorescent wriggling streamers like the tentacles of an octopus. Bundles of light stick out from his spine. As he stretches out the arms, thus forcing the electric fluid outwardly, roaring tongues of fire leap from his fingertips. Objects in his vicinity bristle with rays, emit musical notes, glow, grow hot. He is the center of still more curious actions, which are invisible. At each throb of the electric force myriads of minute projectiles are shot off from him with such velocities as to pass through the adjoining walls. He is in turn being violently bombarded by the surrounding air and dust. He experiences sensations which are indescribable.[9]

HARNESSING THE FALLS

The first attempt to harness the power of Niagara Falls was in 1882, when a local entrepreneur diverted water down a canal to waterwheels powering seven small factories, including flour and pulp mills and a silver-plating plant. Further industrial development nearby was halted by the creation of the Niagara Reservation in 1885.

The following year, civil engineer Thomas Evershed (1817 – 90), who had worked on the Erie Canal, proposed digging a series of canals and tunnels to carry water from Niagara Falls to waterwheels that would be used to power industrial mills and factories.

The intake would be more than a mile above the falls, well out of sight. Plans were made to feed some two hundred waterwheels. These would discharge into a two-and-a-half mile tunnel that would then discharge into the river below the falls. Permission was granted, but the company never managed to raise the capital.

Three years later, Edison planned to electrify Buffalo, 20 miles (32 kilometers) away. However, DC had never been transmitted more than one or two miles. At the time, even Westinghouse was dubious that electricity could be transmitted so far and suggested a complex system of compressed air tubes and cables to convey the power. Plans were drawn up for the construction of an industrial complex next to the Falls, but then came the news that AC power had been transmitted the 109 miles (175 kilometers) from Lauffen to Frankfurt.

CONVERTING KELVIN

The International Niagara Commission, headed by Sir William Thomson, later Lord Kelvin, offered $20,000 for the best plan to harness the power of the Falls. Like Edison, Kelvin was opposed to AC – until he saw it in action at the Columbian Exposition. Then he became an enthusiastic convert. Westinghouse refused to enter at first, as he felt that, to win, he would be handing over $100,000-worth of advice. Of the twenty schemes submitted, fourteen used hydraulics or compressed air. Four involved DC power, one of which was endorsed by Edison. Two used AC. One of them was not fully worked out; the other used the Tesla system manufactured by Westinghouse.

The tunnels were under construction while the battle between AC and DC still raged on. As this was to be a power plant rather than a lighting plant, DC was initially favored – until Tesla's single-phase AC motor proved itself a commercial reality.

REALIZING THE POTENTIAL

In the spring of 1891, the Gold King Mine at Telluride, Colorado approached Westinghouse, asking whether the 320-foot (100-meter) waterfall nearby could be used to power their stamping and crushing mill, otherwise it would have to close. Local wood was exhausted and transporting coal there was too expensive.

Inside Power House No. 2, Niagara Falls.

LORD KELVIN
(1824 – 1907)

Scottish engineer, mathematician and physicist, William Thomson was knighted in 1866 and made a peer as Baron Kelvin of Largs in 1892 for his services to science and engineering.

He helped develop the Second Law of Thermodynamics, the mathematical analysis of electricity and magnetism, the electromagnetic theory of light, the geophysical determination of the age of the Earth and the basics of hydrodynamics. His work on submarine telegraph cables helped make Britain the hub of global communications. It also made him a rich man. He perfected the mariner's compass and worked out the correct value of absolute zero. The units of the absolute temperature scale are named Kelvins in his honor.

With the growth of electric lighting and power in the 1880s, the firm he owned of Glasgow instrument makers added electrical measuring instruments to its production. Meanwhile he became directly involved in numerous electrical projects which ranged from electric traction for trams and trains to the production of hydroelectric power from Niagara Falls and in the Scottish highlands. By the end of his life he had a total of some seventy patents to his name.

Westinghouse installed a single-phase AC generator attached to a waterwheel in a shack next to the falls. Electricity at 3,000 volts traveled up three miles of mountain terrain on copper wires. At the mill, the voltage was stepped down and the current ran a 100-horsepower single-phase Tesla motor. The system continued working, despite blizzards, avalanches, high winds, and electrical storms.

"The aggregate time lost was, by actual count, less than 48 hours during three-fourths of a year," Charles Scott told the *Electrical Engineer* in June 1892. "This success is confirmed in a substantial way by the immediate extension of the plant. A 50 horsepower motor is now being installed at a mill a few miles from the Gold King ... work in this field is fast passing from experimental investigation into practical electrical engineering."[10]

In the same article, Scott described how another Tesla AC generator had been installed at the forty-foot falls on the Willamette River in Oregon, replacing the DC station there that had been destroyed. Electricity was then sent thirteen miles to the central lighting station in Portland.

THE ROTARY CONVERTER

The Cataract Construction Company sent consultants to Pittsburgh to test the new Westinghouse AC generators. They also wanted to see the workings of the new rotary converter that could turn AC into DC, which was important for streetcars that still ran on the old system. Their report said:

The 100-horsepower single-phase AC generator used to drive a stamp mill at the Gold King Mine, Colorado, 1891.

> *A careful examination of the work done in this establishment showed excellent workmanship and correct engineering design in all the machinery examined ... The workmanship is beyond criticism in quality.*[11]

Leading the delegation, Henry Rowland, physics professor from Johns Hopkins University, concluded that Westinghouse had "the greatest experience in the practical use of the alternating system and they seem to control the most important patents."[12]

Visiting the GE plant in Lynn, Massachusetts, consulting engineer Coleman Sellers said that "very considerable change would have to be made to make it mechanically equal."[13]

GE had given up on Edison's DC and were planning to use three-phase AC. Sellers said: "I should incline to the biphase on account of its greater simplicity and its adaptability to a broader field of usefulness."[14]

There were still some doubts about Westinghouse's patents, some of which he had bought in from abroad. But Sellers said: "I am not aware of any claim to ownership in this country of what can stop the owners of the Tesla patents from commanding the market."[15]

THE TESLA CONNECTION

The president of the Cataract Construction Company, Edward Dean Adams, adopted the simple expedient of contacting Tesla, who reassured him of the broad scope of his AC patents in notes scrawled on stationery from the Gerlach Hotel, where he was now staying. Any other company offering multiphase AC generators or AC motors would be infringing the patents he had given to Westinghouse.

Knowing how eager Westinghouse now was to secure the Niagara contract, Tesla wrote to Adams on February 2, 1893:

> *I have not heard from Germany yet, but I have not the slightest doubt that all companies except Helios – who have acquired the rights from my company – will have to stop the manufacture of polyphase motors. Proceedings against the infringers have been taken in the most energetic way by the Helios Co. It is for this reason that our enemies are driven to the single-phase system and rapid changes of opinion.*[16]

On March 12, Tesla assured Adams that the patents held by Thomson-Houston had "absolutely nothing to do with my discovery of the rotating magnetic field and the radically novel features of my system of transmission of power disclosed in my foundation patents of 1888. All the elements shown in the Thomson patent were well known and had been used long before."[17]

Adams then sought Tesla's opinion on the use of DC. Tesla wrote on March 23, explaining "how disadvantageous, if not fatal, to your enterprise such a plan would be, but I do not think it possible that your engineers could consider seriously such a proposition of this kind."

TESLA'S TRUE SIGNIFICANCE

Westinghouse's triumph at the Chicago World's Fair – the largest lighting project ever undertaken at that time – left no one in doubt about who should be awarded the contract. Edward Dean Adams said:

> *The construction of twelve polyphase alternators of a thousand horsepower each and the electrical illuminations of a great White City for the first time in history were great events, but they were overshadowed in real significance by a more important though less spectacular exhibit.*[18]

He was talking about the working model of Tesla's universal AC power system, with its AC generator, transformers, transmission lines, working induction motors, synchronous motor, and Westinghouse's rotary converter that supplied direct current for the streetcar motors.

The reputation of GE had already been besmirched when they were seen to be overcharging in the bids to illuminate the Chicago World's Fair. Then blueprints went missing from the Westinghouse works and were found in GE's Lynn plant, they were accused of industrial espionage.

THERE MAY BE TROUBLE AHEAD

Having studied the rival plans, the Cataract Construction Company informed the competitors that their own electrical consultant, Professor George Forbes, would design the generators. Tesla wrote to Adams saying that he was not the least concerned about this outcome. However, he could not help seeing "difficulty ahead."[19] He could not see how Forbes could avoid infringing his patents on the alternating-current system, let alone several taken out by Westinghouse on improvements.

Nevertheless Forbes went ahead. Then the Cataract Construction Company had the task of looking for a company that could manufacture the generators for them. Westinghouse was approached. He wrote back:

> *We have given several years time to the development of power transmission, and have spent an immense sum of money working out various plans, and we believe we are fully entitled to all the commercial advantages that can accrue to us ... we do not feel that your company can ask us to put that knowledge at your disposal so that you may in any manner use it to our disadvantage.*[20]

Eventually, Westinghouse relented. Again this was the greatest electrical project in the world and he wanted a part of it. It would be another triumph for AC. He sent two engineers to review Professor Forbes' design. They concluded that "electrically it was defective and if built as designed ... would not operate."[21]

It also worked at 16 ⅔ cycles per second. At such a low frequency, it would, at best, produce a flickering light. The frequency was far too low to operate Tesla's polyphase motors and Westinghouse's converter to produce DC. What's more the generating voltage – 22,000 volts – was so high that it would cause possibly insuperable insulation problems.[22]

They eventually compromised on a frequency of 25 cycles per second. However, Westinghouse remained suspicious of Forbes, seeing him as a possible rival in dynamo design. He was sidelined.

THE HYDROELECTRIC REVOLUTION

Westinghouse set about building two of Niagara's 5,000-horsepower generators. They were completely new machines, five times more powerful than those at the World's Fair. The revolutionary nature of the design was spelt out in a report on the first Niagara hydroelectric plant by the Westinghouse Company which said:

> *The switching devices, indicating and measuring instruments, bus-bars and other auxiliary apparatus, have been designed and constructed on lines departing radically from our usual practice. The conditions of the problem presented, especially as regards the amount of power to be dealt with, have been so far beyond all precedent that it has been necessary to devise a considerable amount of new apparatus. The general organization of the cables, switches, and measuring instruments differs materially from anything of the kind hitherto installed elsewhere. Nearly every device used differs from what has hitherto been our standard practice.*[23]

In practice though, they had to be slimmed down to fit on the railroad flatcars that would carry them to Niagara.

The Hydro Electric Power Plant at Niagara Falls.

CHAPTER 12

LIGHTING UP THE WORLD

While the Westinghouse work at Niagara was underway, Edward Dean Adams of the Cataract Construction Company, was busy courting Tesla. He visited him in New York and offered him $100,000 for a controlling interest in fourteen US and foreign patents, along with any future inventions Tesla may come up with. Tesla accepted and, in February 1895, the Nikola Tesla Company was set up. Not only was Tesla working on wireless and remote control, he was putting his mind to cheap refrigeration, the production of liquid air, the manufacture of fertilizers and nitric acid from the air, and artificial intelligence.

Tragedy struck on March 13, 1895, when Tesla's laboratory, then at 33 – 35 South Fifth Avenue, burnt down, destroying all Tesla's equipment. The *New York Sun* wrote:

> *The destruction of Nikola Tesla's workshop with its wonderful contents, is something more than a private calamity. It is a misfortune to the whole world. It is not in any degree an exaggeration to say that the men living at this time who are more important to the human race than this young gentleman can be counted on the fingers of one hand.*[1]

Tesla told *The New York Times*:

> *I am in too much grief to talk. What can I say? The work of half my lifetime, very nearly; all my mechanical instruments and scientific apparatus, that it has taken years to perfect, swept away in a fire that lasted only an hour or two. How can I estimate the loss in mere dollars and cents? Everything is gone. I must begin over again.*[2]

Tesla set about finding a new lab. In the meantime, Edison let him use a workshop in West Orange, and, although uninsured, Tesla was confident that Westinghouse would pay for any new equipment he needed. However, Westinghouse was a hard-headed businessman and billed Tesla. Meanwhile, he announced that he was planning to use Tesla's motors, whose patents he owned, to power locomotives.

Nikola Tesla photographed in 1890 by Napoleon Sarony.

TWIN WIZARDS OF ELECTRICITY

Despite Edison's kindness, the rivalry continued. New York's *Troy Press* asked: "Who is king, Edison or Tesla?"[3] The famous actor Joseph Jefferson even entered the fray, saying: "Edison has been deposed and Tesla has been coronated."

That May, the two men were billed as the "Twin Wizards of Electricity" at the National Electrical Exposition in Philadelphia. Tesla was then on the ascendant as AC had been transmitted along telephone lines for a record-breaking 500 miles (800 kilometers). Tesla was disappointed though as the power transmitted was restricted due to the fear of fire. However, Edison conceded:

> *The most amazing thing about this Exposition is the demonstration of the ability to deliver here an electrical current generated at Niagara Falls. To my mind it solves one of the most important questions associated with electrical development ...*[4]

THE MAN WHO SOLVED THE PROBLEM

Alexander Graham Bell agreed, declaring that he was firmly convinced that this long-distance transmission of electric power was the most important discovery of electric science that had been made for many years. He, with Edison, looking into the future, realized that by means of this discovery, cities and towns remote from the places of electrical generation would be able to obtain the services of this agent for those things for which it is fitted, with great economy, practical safety and far superior convenience than is now possible.

The *Philadelphia Press* hailed Tesla as the man who had "solved the problem." The experiment had shown that electricity generated at Niagara could be used in New York City. Tesla boldly proclaimed his confidence:

> *I am now convinced beyond any question that it is possible to transmit electricity ... to commercial advantage over a distance of five hundred miles at half the cost of generation by steam. I am willing to stake my reputation and my life on this declaration.*[5]

TESLA'S INVENTIVE GENIUS

On July 9, 1895, as work on the "cathedral of power" in Niagara Falls neared its completion *The New York Times* wrote:

> *When Tesla's discoveries were first announced, the electricians of the country and the world at large conceded their novelty and pioneer character, and for several years confined themselves to the forms of apparatus operated by continuous currents, and it was only after the Cataract Construction Company had finally determined that it was by means of Tesla's inventions alone that its operations could be carried out that the actual copying of his work was begun in various quarters. There could be no better evidence of the practical qualities of his inventive genius.*[6]

Construction of the first power station at Niagara had taken five years. It was a headache for investors throughout. The outlay was huge and no-one knew whether it would work as the three-dimensionally imagined plans were all mainly in Tesla's head.

At 7.30 a.m. on August 26, 1895, the turbine of Dynamo No. 2 began to turn, and the AC current generated was sent to power the aluminum smelter of the Pittsburgh Reduction Company, later renamed Alcoa, that had moved its plant up to the falls to take advantage of the cheap electricity. Ironically, aluminum smelting required DC, so after the AC produced by the generator had traveled the short distance to the works, it was passed through three 2,100 horsepower rotary transformers – the largest ever built – to turn it into direct current.[7]

Soon a second generator was up and

running and on September 30, 1895, the board of directors turned up at Power House No. 1 to have their photograph taken. Among them was John Jacob Astor (1864 – 1912) – the richest man to die on the *Titanic* – a friend and supporter of Tesla.[8]

TESLA VISITS NIAGARA

After the opening of Power House No. 1, Tesla was repeatedly invited to visit, but it was not until the summer of 1896 he eventually agreed to come. He first visited Westinghouse in Pittsburgh, then traveled in

POWER HOUSE NO. 1

Also known as the "cathedral of power," Power House No. 1 was designed by Tesla's close friend Stanford White (1853 – 1906). Built in locally quarried limestone with a slate-and-iron roof, it was 200 feet (60 meters) long, 64 feet (20 meters) wide, and 40 feet (12 meters) high. The spartan room housing the generators was lit by tall, arched windows. The turbines were deep in the basement below, and the switchboards sat on a large marble platform. A limestone bridge spanned the inlet canal, carrying the electrical conduits to the transformer house, which was a smaller version of the power house. The plant was named after Edward Dean Adams in 1927 and the original Westinghouse generators – there were eventually ten of them, generating over 35,000 kilowatts – continued working until the plant was shut down in 1961.

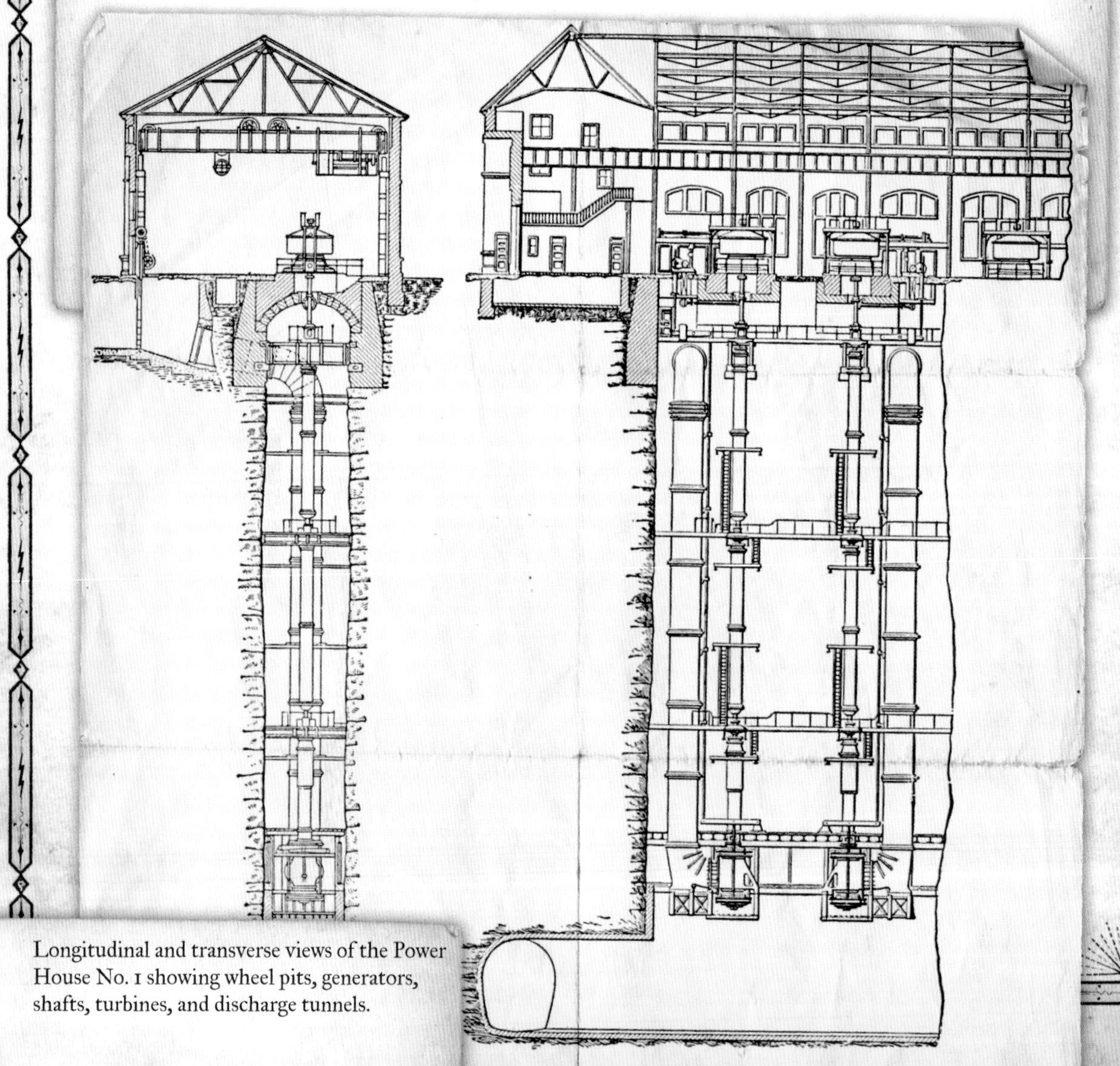

Longitudinal and transverse views of the Power House No. 1 showing wheel pits, generators, shafts, turbines, and discharge tunnels.

Westinghouse's private railroad car with him and Edward Dean Adams to the falls, arriving at 9 a.m. on July 19, 1896.

It was a Sunday morning, so only one dynamo was in operation. Tesla climbed up on the special walkways to inspect it. After studying the machinery on the ground floor, they descended in the elevator to visit the wheel pits below. There, they would see the turbines turning and hear the river water rushing by.

Then they crossed the inlet canal to the transformer house. It was still empty as the transformers were to be supplied by GE. After lunching in the Cataract Hotel overlooking the American Falls, Tesla agreed to speak to the press:

> *I came to Niagara Falls to inspect the great power plant and because I thought the change would bring me needed rest. I have been for some time in poor health, almost worn out.*

Asked what he thought of the power plant, his face lit up:

> *It is all and more than I anticipated it would be. It is fully all that was promised. It is one of the wonders of the century ... a marvel in its completeness and in its superiority of construction ... In its entirety, in connection with the possibilities of the future, the plant and the prospect of future development in electrical science, and the more ordinary uses of electricity, are my ideals. They are what I have long anticipated and have labored, in an insignificant way, to contribute toward bringing about.*[9]

THE CATHEDRAL OF POWER

Looking to the future, as always, Tesla predicted: "The result of this great development of electric power will be that the Falls and Buffalo will reach out their arms and will join each other and become one great city. United, they will form the greatest city in the world."

Tesla admitted that it was his first visit to the falls – and his first visit to the cathedral of power.

Power House No. 1 of the Niagara Falls Power Company.

"I came purposely to see it," he said. "But, and it is a curious thing about me, I cannot stay about big machinery a great while. It affects me very much. The jar of the machinery curiously affects my spine and I cannot stand the strain."

The *Niagara Gazette* described the great man as they witnessed him at the Falls:

Tesla is an idealist, and anyone who has created an ideal of him from the fame that he has won, will not be disappointed in seeing him for the first time. He is fully six feet tall, very dark of complexion, nervous, and wiry. Impressionable maidens would fall in love with him at first sight, but he has no time to think of impressionable maidens. In fact, he has given as his opinion that inventors should never marry. Day and night, he is working away at some deep problems that fascinate him, and anyone that talks with him for only a few minutes will get the impression that science is his only mistress, and that he cares more for her than for money and fame.[10]

Afterwards, Tesla returned to New York, where he threw himself back into his wireless research, fearing that Marconi might steal a march.

THE ELECTRIC FUTURE

When British science fiction writer H.G. Wells visited Power House No. 1, he wrote:

These dynamos and turbines of the Niagara Falls Power Company impressed me far more profoundly than the Cave of the Winds; they are indeed, to my mind, greater and more beautiful than accidental eddying of air beside a downpour. They are will made visible, thought translated into easy and commanding things. They are clean, noiseless, starkly powerful. All the clatter and tumult of the early age of machinery is past and gone here; there is no smoke, no coal grit, no dirt at all. The wheel pit into which one descends has an almost cloistered quiet about its softly humming turbines. These are altogether noble masses of machinery, huge black slumbering monsters, great sleeping tops that engineer irresistible forces in their sleep ... A man goes to and fro quietly in the long, clean hall of the dynamos. There is no clangor, no racket ... All these great things are as silent, as wonderfully made, as the heart in a living body, and stouter and stronger than that ... I fell into a daydream of the coming power of men, and how that power may be used by them.[11]

THE NIAGARA RIVER FLOWS UPHILL

On December 16, 1896, the city of Buffalo signed a contract with the Cataract Power and Conduit Company to supply 10,000 horsepower of electricity on or before June 1, 1897. The first customer was the Buffalo Street Railway Company, who contracted for 1,000 horsepower of DC. Some still had their doubts, but Tesla told the *Western Electrician*.

Its success is certain. The transmission of electricity is one of the simplest of propositions. It is but the application of pronounced and accepted rules which are as firmly established as the air itself.[12]

Westinghouse was also sanguine that they would soon sell 100,000 horsepower from the Niagara plant.

When you think that a single ocean steamship like the Campania uses 25,000 horsepower, it is easy to be seen that there will be no surplus here. All the power here can and will be used.[13]

The company set about erecting a line of wooden transmission poles, similar to those used by telegraph companies. Meanwhile, in November the GE transformers arrived early. They were tested on November 15 and, at one minute past midnight on Monday, November 16, the switches were thrown and the current began to flow.

The 2,200 volts coming from the generator were stepped up by the transformer to 10,700 volts. This was conducted down the twenty-six miles of copper cable. At Buffalo this was stepped down to 440 volts AC, which was fed to a rotary transformer, yielding 550 volts DC.

The *Niagara Falls Gazette* reported: "The turning of a switch in the big powerhouse at Niagara completed a circuit which caused the Niagara River to flow uphill."[14]

After an hour, the current was switched off, while all concerned went off to celebrate, and, the *Buffalo Enquirer* announced the city's renewed prosperity. Already the world's sixth biggest commercial center, it was now about to have the country's cheapest electricity.

THE GREATEST ELECTRICIAN ON EARTH

Tesla had advised against having any celebration for the inauguration of the Niagara Falls power station when it was first opened, but the men running the Cataract Power and Conduit Company were now determined to have a banquet in the Ellicott Club in Buffalo on January 12, 1897. Tesla was to be guest of honor. Three hundred and fifty of America's most prominent scientists and businessmen made the trek there. Each guest was handed a souvenir menu and seating plan, bound with covers made of engraved aluminum, smelted using the company's electric power.

"Such a company never sat down in Buffalo before," said the *Buffalo Morning Post*, "while such an event had never previously been celebrated in the history of the world."[15]

Edison was notable by his absence. Westinghouse was not there either as he disliked all forms of public ceremony.

In a room bathed in electric light, Tesla was introduced as the "greatest electrician on Earth,"[16] and he received a standing ovation. In his speech, he said:

We have many a monument of past ages; we have the palaces and pyramids, the temples of the Greek and the cathedrals of Christendom. In them is exemplified the power of men, the greatness of nations, the love of art and religious devotion. But the monument at Niagara has something of its own, more in accord with our present thoughts and tendencies. It is a monument worthy of our scientific age, a true monument of enlightenment and of peace. It signifies the subjugation of natural forces to the service of man, the discontinuance of barbarous methods, the relieving of millions from want and suffering.[17]

EDISON IS A RELUCTANT CONVERT

Other speakers did not share this lofty view. J.P. Morgan's attorney Francis Lynde Stetson pointed out that New York investors had put up over $6 million to build the Niagara Falls power plant and its transmission system, and they had, so far, not received one cent back in profits or dividends. He told his now-silent audience that the most profitable way to exploit this new form of power was to sell it all to firms that set up in the industrial park around the power station. Nevertheless the company would stand by its agreement to provide Buffalo with 10,000 horsepower of electricity, but not by the deadline of June 1.

He then reminded Tesla that they had just three minutes to catch the train back to New York. Tesla parted with the words: "Let me wish that in no time distant your city will be a worthy neighbor of the great cataract which is one of the great wonders in nature."[18]

The battle of the current was now over. Tesla and Westinghouse had won. While the first one thousand horsepower of electricity reaching Buffalo had been taken by the streetcar company, the local power company already had orders from residents for five thousand more. Within a few years, the number of AC generators at Niagara Falls reached the planned ten, and power lines ran as far as New York City. Broadway was ablaze with lights. It powered streetcars and the subway system. Even Edison's networks converted to alternating current.

55
45
3'
45'
3
46
3'
57
45''
42
41
52
45
38
56
36
3
39
44
44'
44''
44
37

PART FIVE

INVENTING A NEW WORLD

CHAPTER 13

THE GREAT ENTREPRENEUR

With the world irrevocably converted to AC, Westinghouse, Edison and Tesla went their separate ways, though they remained rivals in their different fields. Edison and Tesla continued to vie with one another with their other inventions, while George Westinghouse remained the great entrepreneur, exploiting the ingenuity of others. He had used his industrial might to bring Tesla's greatest innovation to the world, but as a practical man he had had to share the prize with his greatest rival, General Electric, who, now unhampered by Edison's research budget, remained the biggest corporate player in the industry.

While other electrical companies had collapsed during the patent onslaught of the "Seven Years' Incandescent Light Bulb War," Westinghouse survived, and George Westinghouse's spirit of enquiry remained undimmed. Always searching for bigger and better machinery, he became interested in the steam turbine.

This had first been invented by the Swedish scientist Karl Gustaf Patrik de Laval (1845 – 1913), the inventor of the centrifugal cream separator in 1878. He built his first impulse steam turbine in 1882, patenting it the following year. It attained a speed of 42,000 revolutions a minute and, by 1896, had a power plant working.

The New York Edison company imported a 300-kilowatt (400 horsepower) de Laval turbine and Westinghouse began searching around for patents. He came across the work of Englishman Charles Parsons, who had independently developed the steam turbine in 1884. By 1889, he had built some three hundred turbines, running up to 75-kilowatt capacity. Over the next five years, he made a number of turbines of 350- to 500-kilowatt capacity and in 1894, *Turbinia*, the first turbine ship, was launched by the Parsons Marine Steam Turbine Company.[1]

LIFELONG PASSION FOR STEAM

In 1895, Westinghouse bought a license for the manufacture of Parsons' turbines in the US. Steam was a lifelong passion. His first patent in October 31, 1865, had been for a rotary steam engine. The Westinghouse Electric Company installed three 300-kilowatt machines in the Westinghouse Air Brake Company in Wilmerding, Pennsylvania, in 1898. The following year a 1,500-kilowatt turbo-generator, running at 1,200 revolutions per minute was installed in the power plant of the Hartford Electric Light Company.

Scaling up a steam turbine caused problems because of the small clearances required to maintain efficiency. These were solved in England by building the turbine in two sections, one using high-pressure and the other low-pressure steam. Westinghouse felt this caused an unnecessary increase in the size and cost. Instead he solved the problem by devising the "single-double-flow" type.[2]

THE TURBO-GENERATOR

As soon as he had perfected the steam turbine, Westinghouse saw that it was the perfect device for driving a Tesla polyphase generator for heavy power generation. He came up with the idea of putting the steam turbine and the electrical generator together in one unit.

The first turbo-generators he built for the Air Brake Company, produced 440 volts, at 60 cycles per second, polyphase. It spun at 3,600 revolutions per minute and the armature and end winding began to distort. So in 1899, it was decided to use the type of rotating field developed by Tesla. The rotor had many relatively small slots to subdivide the field winding into a large number of small coils, so that each could be supported without unduly crushing the insulation or becoming displaced. This developed into the "parallel-slot" construction, which was used by the company for many years and made Westinghouse the pacemaker in the race toward higher speeds.[3]

The first 3,600-revolutions-per-minute, parallel-slot rotor for the 60-cycle generator made such a frightful noise that Westinghouse sent his engineers away. Modifications were made and the improved design was in production for ten years.

With other improvements such as artificial cooling, the power of the 3,600-revolutions-per-minute turbo-generator was increased from 400 kilowatts up to 6,250 kilowatts and the twenty-five-cycle, two-pole, 1,500-revolutions-per-minute machine increased

BUILDING THE *TURBINIA*

Built in light steel at Wallsend-on-Tyne in the north-east of England, *Turbinia* was launched on August 2, 1894. She was 100 feet (30 meters) long with a draft of 3 feet (1 meter) and a beam of 9 feet (3 meters), and powered by a 1,000 horsepower engine capable of 2,000 revolutions per minute, driving a single two-blade propeller.

Trials were disappointing. Propeller slip – that is, the efficiency of propulsion – was 50 percent, caused by cavitation. The single propeller was replaced with three shafts each carrying three propellers, making nine in all. With a top speed of over 34 knots (39 mph or 63 km/h), the *Turbinia* became the fastest boat in the world. Ten years later the British navy decided that all future ships would be turbine-powered.

The *Turbinia* is now on display at the Discovery Museum in Newcastle, England, and is designated an International Landmark by the American Society of Mechanical Engineers.

Steam turbine-powered *Turbinia* at speed, photographed in *c.* 1897.

its output from 750 kilowatts up to 10,000 kilowatts, or 13,400 horsepower, within a few years.

POWERING THE NEW YORK SUBWAY

By 1902, 6,000 kilowatt units were being built to power the New York subway. The outside diameter of the armature frames was 42 feet (12 meters). There was not enough room in the existing shops in East Pittsburgh to assemble them. A new aisle had to be built with traveling cranes, 44 feet (13 meters) above them. They were so big that they had to be delivered to Manhattan in four pieces and assembled there. The output still had to be DC.

Westinghouse was still determined to oust DC. He put aside his dislike of public speaking and agreed to serve as chairman for the International Railway Congress in Washington, D.C., in May 1905, attended by over a thousand delegates from forty-eight nations. He used it to showcase his new AC locomotive.

When the Congress was over, Westinghouse took three hundred railroad men back to Pittsburgh on a special train. There he gave a demonstration of the first AC locomotive, which he maintained was the "first real main-line locomotive."[4]

By June 1907, he had converted the New York, New Haven, and Hartford Railroad to the single-phase system. According to the corporate history, this was the "consummation of Westinghouse's great career."[5]

SINGLE-DOUBLE-FLOW

The vice-president of Westinghouse Electric & Manufacturing Company Herbert T. Herr explained how single-double-flow worked:

> *The turbine is essentially a machine for developing large powers, and it reaches its maximum economy with large capacity. It is essentially different from the ordinary steam engine in that it converts the energy in steam into mechanical work by utilizing the velocity resulting from the steam expansion, either by action or reaction of a steam jet on the blades, as opposed to the conversion of steam into energy in reciprocating engines by direct pressure of the steam on a piston.*
>
> *Multiple stages become necessary in the turbine to fractionally extract the energy of steam in its expansion from boiler pressure to the condenser because it is impossible in mechanics of engineering, as now known, to provide materials which would stand the stresses and speed necessary to extract in one stage efficiently the energy of a jet of steam expanding from 200 pounds pressure to 29 inches vacuum, as the steam speed under these conditions would be 4,300 feet per second.*
>
> *In turbines of large capacity, on account of the large volumes of steam to be handled in the low-pressure stages, we again encounter the difficulty of materials in mechanical construction to efficiently handle them through the blading, and it is therefore necessary to divide the steam in such cases, and flow half of it through blading of half the area which would be required if the turbine were single-flow. In other words, by double-flowing you can double the capacity of the machine.*
>
> *While the double-flow turbine is an old construction, Mr. Westinghouse conceived the idea of using a single-flow construction in the upper ranges of the turbine and then, in the same cylinder casing, dividing the steam and passing it through two independent low-pressure portions; hence the name single-double-flow turbine. Of course, the whole turbine could be made double-flow, but it would mean a spindle of twice the length of a single-flow turbine, and by double-flowing only the low-pressure portion the machine is shortened and cheapened.*[6]

An Engineering Wonders cigarette card, 1927, featuring the Baldwin-Westinghouse electric locomotive.

REOPENING THE RIVALRY

The development of the turbo-generator reopened the rivalry between Westinghouse and General Electric. Westinghouse used a horizontal shaft, while General Electric favored a vertical one. But as speeds increased, constructing and operating the vertical type became increasingly difficult and, in the end, GE had to change from vertical to horizontal.

Turbo-generators were much smaller than the older units powered by reciprocal engines. It was another of Westinghouse's revolutionary contributions to engineering. Westinghouse's biographer Henry Prout, said:

> *This obsolescence of the engine-type alternator was almost pitiful. Here was a branch of heavy engineering, built up at great cost and backed by years of experience. In the coming of the turbo-generator this experience was mostly thrown away, for the engineering required in the turbo-generator work was so radically different from that of the engine-type generator that the designers had to start practically anew and build up entirely new experiences at enormous expense and through years of effort.*[7]

MARINE PROPULSION

Electrical generators are efficient when run at high propulsion speed, but when a ship's propeller is turned too fast the result is slip and cavitation. Gearing was needed between the turbine and the propeller shaft. This presented difficulties on board a ship as any deflection or misalignment would cause excessive wear.

Nevertheless, a system was devised which Westinghouse called the floating-frame gear, and an engine capable of developing 5,000 horsepower costing $75,000 was installed on the US collier *Neptune*. The Westinghouse steam turbine was then adopted for US Navy war ships.

While the Westinghouse Machine Company went into receivership, Westinghouse continued his experimentation, building a concrete tank 18 feet (5.5 meters) in diameter and 6 feet (2 meters) deep, containing a shaft driven by a 500-horsepower turbine, to measure slip and cavitation.

He investigated the possibilities of lubricating the propeller blades and the hull of a ship with a film of air. Tests were made on an electric launch on Laurel Lake.

"To produce the required quantity of air I have had made a rotary blower in which are incorporated, in a new manner, details which have been in use some years," Westinghouse said in a letter to Lord Kelvin. "I find a sheet of air one-half inch thick can be paid out next to the hull, from slots, as fast as the ship moves. The *Lusitania,* for instance, would require less than 600 horsepower (to deliver the air), and this should so reduce the skin friction as to greatly affect the speed. I have discovered, however, that the air in the water will necessitate the use of a special propeller, to avoid cavitation, which I am having made for trial on the launch."[8]

Lord Kelvin replied: "I do not think it possible that good results can be got by air lubrication of the hull of a ship or of the blades of a propeller. Experiments on a small scale on your electrical launch might seem to promise good results, but I feel perfectly sure that it would be impossible to get good results on the large scale of a ship at sea."[9]

Nevertheless, Westinghouse continued his experimenting, throughout the time some of his other companies were going into receivership.

MAURICE LEBLANC'S CONDENSER

At one stage, Westinghouse thought that gas engines might supplant steam engines for generating electricity, but changed his views when he discovered that the use of high-pressure superheated steam with a high vacuum greatly increased the efficiency of the steam turbine.

A French physicist and engineer named Maurice Leblanc had patented an improved air pump that could be used in connection with existing types of condensers. The mechanism was relatively lightweight, cheap to manufacture, and simple in construction and operation. It also created a considerably higher vacuum than was obtainable with any other type of air pump in use. The increased vacuum reduced the steam consumption of turbines so GE simply helped themselves.

"About 1897, the owners of my patents started a suit for infringement against the General Electric Company," said Leblanc. "This suit took on Homeric proportions – the defense was as vigorous as the attack and it was becoming a celebrated case."[10]

Then, in 1901, Westinghouse was in Europe and resolved the situation. Leblanc retold the story of his meeting with Westinghouse:

I was stopped on the Boulevard by an unknown person, who addressed me in the following terms: "Mr. George Westinghouse, who is now in Paris, leaves for London in two hours; he wishes to see you immediately, and has commissioned me to find you and to take you to him, dead or alive."

I replied: "Then you mean to effect an abduction or to kidnap me. Unfortunately this can no longer be regarded as the abduction of a minor. Well then, kidnap me, I have no objection."

He conducted me to the Rue de l'Arcade, where, for the first time, I saw the great engineer, who said to me: "So it is you who have sworn to make the fortune of all the lawyers in America. Can we come to terms?"

I replied: "I ask for nothing better, and probably my associates will do likewise."

That was all for that day. But I had been greatly struck with the great bearing of the man and his easy good humor. Some months later he bought for the Westinghouse and General Electric Companies my patents, and the inventor into the bargain, whom he appointed consulting engineer to the Société Anonyme Westinghouse in France. That was the starting point of a cooperation of which I shall always be proud, my first impression being duly confirmed. He was before all things a perfect gentleman and a great-hearted man, and he was himself a mechanician beyond compare.[11]

WESTINGHOUSE LOSES CONTROL

Although the introduction of the turbo-generator hugely boosted Westinghouse's sales, the corporation was hit by the banking crisis of 1907. Then Westinghouse Electric went into receivership.

On the day the receivership was announced, *The New York Times* expressed its sympathy:

> *He has done so much for the splendid industrial evolution of this country. His enterprises have been so varied and so important, he has carried the name and fame of American invention and development in the application of novel scientific principles over such wide areas and everywhere has won for his nation such admiration, confidence, and respect, that he presents himself, in a way, as an American institution in who we have patriotic pride.*[12]

Five thousand employees put up $600,000 to help out. It was not enough. Westinghouse needed $4 million to cover his immediate debts. When the electric company emerged from bankruptcy the following year Westinghouse had lost control. By later in 1910, he had been forced off the board.

"The loss of the Electric Company was to Mr. Westinghouse a disappointment from which he never recovered," said his public relations officer Ernest Heinrichs. "There is no doubt that it broke his spirit."[13]

In June 1912, the AIEE awarded him the Edison Medal for "meritorious achievements in connection with the development of the alternating-current system." The irony was not lost on him. Already his health was failing. He died on March 12, 1914. At the time, some fifty thousand people worked for the companies he had created. They were worth $200 million, while his own fortune amounted to $50 million.

THE EDISON MEDAL

The Edison Medal was created in 1904 by a group of Edison's friends and associates as an annual award to be given to a living electrician for "meritorious achievement in electrical science and art." In 1909, the American Institute of Electrical Engineers agreed to present it as their highest award. The first recipient was Tesla's rival Elihu Thomson. George Westinghouse, Nikola Tesla, and Alexander Graham Bell also received the award. After the AIEE merged with the Institute of Radio Engineers, the medal was presented by the Institute of Electrical and Electronics Engineers.

LEBLANC ON WESTINGHOUSE

After Westinghouse's death in 1914, Maurice Leblanc said of him;

> *He inspired me not only with great admiration but also with a warm affection, which I believe he returned to some extent. He was the best and most steadfast of friends. I could obtain witnesses amongst all his old co-workers whom I knew and whose fortunes he had made. All adored as much as they esteemed him. His vigor and his power of work were extraordinary. He never took any rest. Starting with next to nothing, he became one of the greatest industrial captains in the world. He fell in action, crushed like a Titan, on the eve of the Great War, which he had long foreseen. In fact, in 1903, he said to me that the first United States war would be against the insupportable Germans. His talent as an organizer would have been of the very greatest service, and this for us is a further cause to regret his premature end. George Westinghouse was a great American, and no man had a greater regard for his country. He lived like the type of modern inventors and great realizers. His memory will always be green in the hearts of those who surrounded him and all who loved him.*[14]

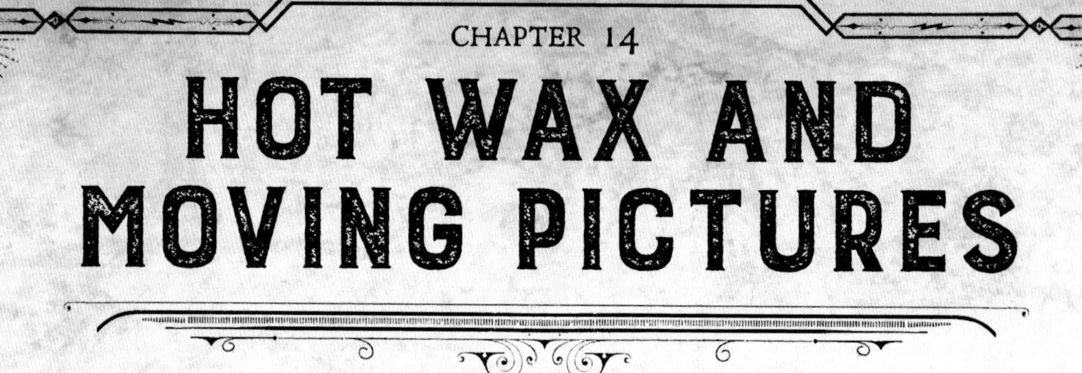

CHAPTER 14

HOT WAX AND MOVING PICTURES

Although Edison had lost the battle of the currents, his enthusiasm for new projects was undimmed. Believing that America was running out of iron ore, he decided that he could extract more out of played-out East Coast mines using powerful magnets. To do that, he would have to build huge machines to pulverize rock into a fine powder.

He set about doing this, in the process building a fully automated factory that required no human labor – except for the staff of two hundred engineers who fixed the machines when they broke down. It cost $2 million and was known as "Edison's Folly."

Eventually, in 1898, his Ogdensburg plant in New Jersey was crushing four- and five-ton rocks and selling the magnetite he extracted to Bethlehem Steel. But while he had cut his costs from $7 a ton to $4.75, the great Mesabi iron range had opened in Northern Minnesota. Suddenly iron cost only $3 a ton. By 1899, Edison had to close the iron mill for good. It was not a total disaster though. His friend Henry Ford (1863 – 1947) picked up his ideas for his assembly line from Edison's plant, and *Iron Age* magazine proclaimed that Edison was twenty-five years ahead of his time.

Edison himself went on to use his stone crushing equipment to make cement. His idea was to use the cement to make prefabricated houses. He constructed several in West Orange at a cost of $1,200 each. Again he was ahead of his time as prefabricated houses did not catch on until decades later. Nevertheless, by 1905, Edison's cement plant was the fifth largest in the country.[1]

(Left) Thomas Edison with a model of a concrete house. (Right) An Edison concrete house under construction, 1911.

PHONOGRAPHIC IMPROVEMENTS

Edison continued working at his West Orange labs, seeking ways to improve his phonograph. His machines began to make money when they were good enough to reproduce music. Using the "Gold-Molded" process he developed in 1902, it proved possible to make a metal mold from a wax master. Then fresh hot wax copies could be made from the mold.

Artists were keen to record for Edison and thousands of titles were issued. In 1908, Edison introduced the Amberol cylinder, with the grooves closer together. This gave four-minutes of playing time rather than two. In 1912, Blue Amberol took over where the cylinder was made from celluloid instead of wax.[2]

At the same time, Edison started Edison Disc Phonograph. Although discs did not inherently provide greater audio fidelity, the manufacturing process was easier. Discs can simply be stamped out. However, Edison continued making cylinders until the fall of 1929.

A tabletop phonograph with blue celluloid cylinder, around 1920.

MOVING PICTURES

Edison was a pioneer in the world of the movies. As with the incandescent lamp, others were at work in the field, but it was Edison who turned the theory into reality and introduced a practical device to the market place.

In France, the pioneer of color photography Louis Arthur Ducos de Hauron (1837 – 1920) had filed a patent describing a means of making motion pictures in 1864. But Ducos never got the idea beyond paper. His ideas for color photography were taken up by the Lumière brothers, who beat Edison to the punch with their Cinematograph, holding the first public screening of a motion picture in 1895, in Paris.

THE TROTTING HORSE

In America, in 1872, British photographer Eadweard Muybridge (1830 – 1904) was also experimenting with motion photography at the behest of Senator Leland Stanford, a California railroad magnate and horse-lover, who asserted, contrary to popular belief that, at some point in its gait, a trotting horse had all four hoofs off the ground.

Muybridge set up a bank of cameras whose shutters were triggered as the horse trotted past. By printing the photographs on a rotating glass disk, it was possible to project them onto a screen in rapid succession, giving the illusion of continuous motion. Describing the results the *Scientific American* of June 5, 1880 said:

While the separate photographs had shown the successive positions of a trotting or running horse in making a single stride, the Zoogyroscope threw upon the screen apparently the living animal. Nothing was wanting but the clatter of hoofs upon the turf, and an occasional breath of steam from

Galloping horse film sequence by Eadweard Muybridge.

the nostrils, to make the spectator believe that he had before him genuine flesh-and-blood steeds. In the views of hurdle-leaping, the simulation was still more admirable, even to the motion of the tail as the animal gathered for the jump, the raising of his head, all were there. Views of an ox trotting, a wild bull on the charge, greyhounds and deer running and birds flying in mid-air were shown, also athletes in various positions.[3]

Using a chronophotographic gun. Illustration from *La Nature* magazine, 1882.

Two years later Professor E.J. Marey of the French Academy, another expert on animal movement, made further advances using dry plates. With a single camera, he was able to obtain twelve photographs on successive plates in one second. The camera was in the form of a gun, so that it could follow a moving object, putting each image in the center of the plate. However, like Muybridge, he was limited to a short cycle of movements of a single object, repeated over and over again.

BRINGING BACK THE DEAD

For Edison, none of this was good enough. He said:

In the year 1887, the idea occurred to me that it was possible to devise an instrument which should do for the eye what the phonograph does for the ear, and that by a combination of the two, all motion and sound could be recorded and reproduced simultaneously. This idea, the germ of which came from the little toy called the Zoetrope and the work of Muybridge, Marey, and others, has now been accomplished, so that every change of facial expression can be recorded and reproduced life-size. The kinetoscope is only a small model illustrating the present stage of the progress, but with each succeeding month, new possibilities are brought into view. I believe that in coming years, by my own work and that of Dickson, Muybridge, Marey, and others who will doubtless enter the field, grand opera can

EADWEARD MUYBRIDGE
(1830 – 1904)

Born in England, Muybridge immigrated to the US at age twenty. He took up photography, selling prints from the bookstore he opened in San Francisco. He also demonstrated ingenuity with mechanical devices, taking out patents for improvements in printing and machinery for washing clothes.

In 1866, he went into the photography business full-time, building an extensive catalog of views of the American West, recording the emerging San Francisco, and accepting various positions as photographer for government boards and expeditions.

In 1872, the president of the Central Pacific Railroad, and racehorse owner, Leland Stanford, hired Muybridge to prove his theory that a trotting horse has all four legs off the ground at one point. Muybridge's experiments were interrupted when he shot and killed his wife's lover. He was acquitted.

Developing a shutter that gave an exposure of 1/500th of a second, he set up a bank of cameras that proved Stanford's contention in 1877. To illustrate his lectures on animal locomotion, Muybridge developed the zoopraxiscope, a lantern that projected images in rapid succession on a screen, giving the impression of movement. This was a sensation at the World's Fair at Chicago in 1893.

Under the auspices of the University of Pennsylvania, he famously produced studies of human figures, clothed and naked, performing various activities. He returned to his birthplace in England in 1894 where he died, ten years later.

> *be given at the Metropolitan Opera House at New York without any material change from the original, and with artists and musicians long since dead.*[4]

In February 1888, Muybridge visited Edison in New Jersey and proposed combining his zoopraxiscope with Edison's phonograph, so that sounds and images could be played concurrently. But, Edison was already working on a way to secure photographs reduced microscopically along a spiral around a cylinder. The result would be viewed with a magnifying glass.

This proved impractical as the emulsions employed were too coarse to give a sharp image at microscopic size. If the pictures were any heavier, it would be impossible to stop and start them quickly enough. So he put William Dickson on the job. Nevertheless Edison continued work on the stop-go action of the cylinder.[5]

Dickson got the best results by wrapping photographic paper around an aluminum cylinder. However, because of the curvature of the cylinder, the top and bottom were out of focus. Instead, Dickson introduced a cylinder with flat sides.[6]

READING MEN'S MINDS

While Dickson continued work on the device, Edison went to the Paris Exposition in 1889. He was asked there whether a machine could be made to read men's minds.

"Such a machine is possible," he said after a few minutes reflection. "But just think if it were invented, every man would flee his neighbor, fly for his life to any shelter."[7]

He met a number of distinguished photographers at a dinner celebrating the achievements of Louis-Jacques-Mandé Daguerre (1787 – 1851), the inventor of the daguerreotype – the first practical photographic process.

WILLIAM KENNEDY DICKSON (1860 – 1935)

Scottish inventor William Dickson was born in Brittany, France, and immigrated to the US in 1879. He went to work for Edison as the company's official photographer. When Edison filed the first caveat on the Kinetoscope, he gave Dickson the task of turning the idea into reality. The first Kinetoscope was shown at the Brooklyn Institute of Arts and Sciences in May 1893.

His team also devised the Kinetograph, a motion-picture camera. He appeared in *Dickson Greeting*, a three-second movie showing him passed his hat between his two hands made in 1891 in Edison's Black Maria studio.

He left Edison in 1895, to set up the American Mutoscope Company which made its own peep-show machines and short movies that often showed young women undressing or posing as artist's nude models.

Film still from *Dickson Greeting* (1891), the oldest American film shown to a public audience.

He visited Marey's studio where he saw Marey's camera that "was capable of producing sixty frames per second on a roll of paper-based photographic film." [8] Edison had mentioned using a continuous strip of film in his original caveat, but had later dismissed it because of mechanical difficulties.

Marey's method reminded Edison of his automatic telegraph that had sprocket wheels to feed the paper and a relay to control the speed. This would be incorporated in the new Kinetograph. The perforations Edison added to the sides would become an industry standard and are known as "American perforations." [9]

MARVELOUS LETHAL INSTRUMENTS

While Edison had been away, Dickson had been working with a tachyscope, invented by the German Ottomar Anschütz (1846 – 1907). Dickson managed to synchronize a voice on a phonograph with the image of the face of a person who appeared to be talking.

This was leaked to the press. In February, 1890, London's *Pall Mall Gazette* reported:

> *Mr. Edison has added a new horror to existence. He attaches an instantaneous photographic camera to his phonograph, plants his two lethal instruments right in front of an orator, and sets them to work. The phonograph records every syllable that fall from his lips, every 'hem' and 'er' and mispronunciation ...* [10]

Then George Eastman (1854 – 1932) – inventor of the Kodak, the first mass-market camera – came up with flexible film that could be rolled around a spool. When Edison first saw it, he exclaimed: "That's it – we've got it – now work like hell." [11]

Throughout the winter of 1890 and the spring of 1891, Edison and his men perfected the Kinetograph using Marey's reel-to-reel method. They abandoned attempts to combine the camera with a phonograph as they proved too hard to synchronize. Edison also gave up on trying to project the image on a screen as this accentuated the jerky nature of the film. [12]

MAKING THAT FINAL STEP COUNT

Edison demonstrated his two new marvels at the West Orange Laboratory for members of the National Federation of Women's Clubs who had come out to New Jersey to dine with his young wife, Mina. Reporters from the *New York Herald* and *New York Sun* were also present. Edison said that the devices would not be launched commercially for another two years. Nevertheless, in June 1891, he gave *Harper's Weekly* an exclusive that said Edison's new invention "performs the same service in recording and then reproducing motion which the phonograph performs in recording and reproducing sound."

Edison modestly told the magazine: "All I have done is to perfect what had been attempted before ... It's just that one step that I have taken." [13]

"Yes, that is all," said the journalist interviewing him, George Parsons Lathrop, son-in-law of Nathaniel Hawthorne and Edison's future biographer. "But in science and invention the clever old maxim does not hold true that 'it is the first step which costs.' Not the first, but the last – the conclusive and triumphant step – is the one that costs and that counts." [14]

The following month, Edison patented the Kinetograph and the Kinetoscope in the US, but he saved himself $150 by not taking out foreign patents. As a result, manufacturers in Europe could make their own, file patents in Europe, and export their machines to the US.

THE TACHYSCOPE

German photographer and inventor Ottomar Anschütz developed a shutter that gave an exposure of 1/1000th of a second. He captured motion using banks of twelve, later twenty-four, cameras. The images were viewed on his version of a zoetrope where the pictures were mounted round the inside of cylinder with a viewing slit between each.

This was developed into the Tachyscope, where twenty-four glass diapositives were mounted on a disk that was cranked and viewed by the intermittent illumination of a spark of a Geissler tube.

By 1891, a motor-driven version was being manufactured by Siemens & Halske, both for home use and as a coin-operated public attraction. Some 34,000 people paid to see it at the Berlin Exhibition Park in the summer of 1892.

Two years later the Projecting Electrotachyscope was used to screen shows in a three-hundred-seat hall in Berlin. Admission was one mark and Anschütz became photographic advisor to the Kaiser.

Ottomar Anschütz's electro-tachyscope. Illustration from *Scientific American* (1889).

THE LUMIÈRE BROTHERS

Auguste Lumière (1862 – 1954) and his brother Louis (1864 – 1948) were the sons of painter turned photographer, Antoine Lumière, and worked for him. At the age of eighteen, Louis developed a commercial method of producing photographic plates. By 1894, their factory was turning out fifteen million plates a year.

That year, father Antoine was invited to a demonstration of Edison's Kinetoscope in Paris. His description inspired his sons who set about work on the cinématographe uninhibited by Edison who had not taken out European patents on his device. They copied his idea of using sprocket-wound film, though slowed the rate from Edison's forty-six frames a second to sixteen.

They combined this with the *Théâtre Optique* of Charles Émile Reynaud (1844 – 1918), who projected the result on a screen, showing the first animated film in public in 1888. The cinématographe worked both as a camera and a projector. In 1894, they shot their first film, *La Sortie des Usines Lumière à Lyon*, showing workers leaving the Lumière factory. It ran for approximately fifty seconds.

The first private screening took place before an audience of two hundred at the Society for the Development of the National Industry on March 22, 1895. The first public screening, when admission was charged, took place at Salon Indien of the Grand Café in Paris on December 28, 1895.

EDISON'S
THE VITASCO
COPYRIGHTED 1896 RAFF & GAMMON.

GREATEST
MARVEL

OPE

"Wonderful is The Vitascope. Pictures life size and full of color. Makes a thrilling show."
NEWYORK HERALD, April 24, '96.

Edison's Greatest Marvel: The Vitascope.

CHAPTER 15

THE WIZARD'S LATEST TRICKS

After showing the Kinetoscope at the World's Fair in Chicago in 1893, Edison realized that he would need films to show on them when they went on sale. William Dickson built an odd-shaped structure at the laboratory. It was covered in tar-paper. The roof could be opened to let in the sun and the whole edifice could be rotated to take full advantage of the lights. Painted black inside and out, it was the world's first purpose-built movie studio and was nicknamed the "Black Maria."

One of the first short pictures to be shot showed Dickson's assistant Fred Ott taking a pinch of snuff and sneezing. Stills from the movie were used to illustrate another article in *Harper's* and *Fred Ott's Sneeze* was the first motion picture to be copyrighted in the US.

Over the next three years, as work continued to improve the Kinetograph, boxers, Gayety girls, Japanese knife throwers, French ballet dancers, strongman Eugene Sandow, Spanish dancer Carmencita, and Buffalo Bill Cody were filmed in the Black Maria. And animal trainers brought dogs, monkeys, bears, and lions along to perform.

In 1896, a film emerging from Edison's Black Maria provoked outrage. Actors John Rise and May Irwin re-enacted for the Kinetograph, the final scene from their Broadway musical, *The Widow Jones*, where they kiss. Called *The Kiss*, it was just eighteen seconds long and the kiss itself – a mere peck – is over in the blink of an eye. Nevertheless moralists filled newspaper columns with demands for movie censorship.[1]

THE MOVIE HOUSE

Housing ten of the "Wizard's Latest Invention,"[2] the first "Kinetoscope Parlor"[3] opened in an old shoe store at 1155 Broadway in New York City on April 14, 1894. Charging twenty-five cents per person, it took $120 the first day. Ten more were sent to Chicago and five to Atlantic City. Edison was charging $250 a machine and soon Kinetoscope Parlors were opening in Europe.

Grey and Otway Latham, who were running a parlor at 83 Nassau Street, New York, approached Edison with a promising screen projector. Always reluctant to embrace the ideas of others too quickly, Edison rejected it out of hand:

Edison's Black Maria film studio, 1893.

If we make the screen machine that you are asking for, it will spoil everything. We are making these peep-show machines and selling a lot of them at a good profit. If we put out a screen machine there will be a use for maybe about ten of them in the whole United States. With that many screen machines you could show pictures to everyone in the country – and then it would be done. Let's not kill the goose that lays the golden egg.[4]

But Dickson made a secret deal with the Latham brothers. They formed the Lambda Company and he built them a camera to go with their projector. When Edison found out, Dickson was fired. Edison then told the *New York Sun* that he was happy that people were making improvements to his invention, but if they called it anything other than the Kinetoscope he would sue.[5]

THE VITASCOPE

In the end, Edison realized that there was a demand for a projector and on April 23, 1896, he showed "Edison's Latest Marvel, the Vitascope" at Koster and Bial's Music Hall in New York. This projected the moving image onto a screen.[6] Edison was not its inventor. Rather his Kinetoscope Company had bought in the patents from Thomas Armat in an attempt to revive its falling sales. His laboratory set about improving the technology.

After three months, Edison made improvements in the mechanism that controlled the movement of the film. By July, he was marketing the "Projecting Kinetoscope" or "Edison Projectoscope."[7]

At first, audiences were delighted by moving pictures of dancing girls, boxers, speeding trains, and crashing waves. Then in

THE PHANTOSCOPE

In 1891, Charles Francis Jenkins (1867 – 1934) began experimenting with movie film which had been developed by Eastman Kodak in 1889. He developed a movie projector that he called the Phantoscope in 1894, showing a hand-colored film of a vaudeville dancer to an audience in his cousin's jewelry store in downtown Richmond, Virginia. He had shot the film himself in the backyard of his Washington boarding house.

With a classmate from the Bliss Electrical School in Washington, D.C., named Thomas Armat (1866 – 1948) he filed patents for a modified version on July 20, 1897. Jenkins sold his interest to Armat who, in turn, sold the rights to Edison, who marketed the invention as the Vitascope.

Jenkins went on to work on television, giving the first public demonstration of "transmitting pictures by wireless" in 1925. However, his mechanical system was eventually superseded by the electronic version.

The Phantoscope movie projector.

1904, Edison's studio produced *The Great Train Robbery*, the first motion picture with a story. It was twelve minutes long and featured action indoors and out. Others followed suit. By 1909, there were eight thousand movie theaters and nickelodeon parlors in the US.[8]

Edison had failed to file patents in Europe and other companies freely borrowed the technology. However, in 1907, a federal court upheld his original patents of 1891, granting him royalties on any motion-picture camera or projector that resembled his originals.[9]

His rivals were forced to form the Motion Picture Patents Company, guaranteeing Edison fees of $1 million a year. With the Edison Manufacturing Company making movie equipment and films, and the Phonograph Works and the National Phonographic Company, Edison now dominated the entertainment industry.

The Motion Picture Patents Company collected a weekly fee from theaters using their equipment. They formed a cartel that refused to sell movies to distributors who took films from other independent movie producers, and cut a deal with George Eastman, so he would only sell raw stock film to the Patents Company. It was broken up by the US Supreme Court in 1917, when independent distributor turned producer, William Fox, sued under anti-trust laws.[10] Fox's Fox Film Corporation eventually became Twentieth-Century Fox.

THE STORAGE BATTERY

Edison had not lost his interest in electricity. He set about developing a storage battery for use in automobiles. In 1900, the only one available was an acid and lead battery that was heavy and corrosive. This gave Edison the opportunity to return to his first love – chemistry.

He employed over ninety people, including leading physicists and chemists, on the quest to find the right electrolyte. His first attempt was a failure. Battery performance dropped off after a year and they began to leak. He withdrew the battery from sale and bought back those that had already been sold.

Edison continued to experiment with his traditional trial-and-error method. Eventually he discovered that the addition of lithium boosted the capacity of the battery. No one knew why until the 1960s.

By 1911, as research continued, Edison re-organized his new companies as Thomas A. Edison, Incorporated with a centralized management.

When World War I began in Europe in 1914, supplies of the carbolic acid that were needed in the phonographic industry were embargoed. Edison set about finding ways to make it and set up his own plant.[11]

But in December 1914, his West Orange Laboratories and thirteen surrounding buildings burnt to the ground at an estimated loss to Edison Incorporated of nearly a million dollars.

"I am sixty-seven," said Edison, "but I am not too old to make a fresh start."[12]

Henry Ford lent him $750,000 while Edison went to work in a leased building nearby with some of the equipment they had managed to rescue. With further bank loans, Edison largely rebuilt his laboratory and the surrounding businesses in three weeks.

By January 1915, employees were hard at work producing his improved disk phonograph. Fearing that the US would be dragged into World War I, Edison accepted a post on the Naval Consulting Board in July that year. He was already supplying the US Navy with storage batteries for its submarines.

In June 1923, *The New York Times* recorded that Edison was worth over $15 billion to the US economy – the equivalent of $3 trillion today, that's not far short of the revenue of the US Government. The report said that it was "within 20 per cent of equaling the value of all the gold dug from the mines of the earth since America was discovered." He was also reckoned to be responsible for the employment of 1,500,000 men.[13]

REPLACEMENT FOR RUBBER

Although Edison was in his mid-seventies, he did not stop work. However he did not often attend the West Orange laboratories, preferring to work in a laboratory in his New Jersey home, or in his country estate in Fort Myers, Florida, where he spent his winters.

With $90,000 each from Ford and Harvey Firestone he worked on a replacement for imported rubber, experimenting with native plants that might yield appropriate forms of latex and synthetic substitutes. Over two years, he collected over seventeen thousand plants. His experiments showed that around two hundred of them contain rubber latex. Edison then began a crossbreeding program with goldenrod, a flowering plant of the aster family native to North America. He produced a variety 12-feet (3.7 meters) tall that yielded 12 percent rubber. Firestone used the latex to make a set of tires that he put on the Model T that Ford had given Edison. He then set about trying to make a synthetic version.[14]

HENRY FORD AND THOMAS EDISON

Henry Ford (1863 – 1947) was chief engineer at the Detroit Edison Company in 1893. As his job was to keep the city's lights on twenty-four hours a day, he was on call at all times, but had no fixed hours. This gave him free time to work on his first horseless carriage, which he called the "Quadricycle." It had a gasoline engine mounted on a buggy frame with four bicycle wheels and was completed in 1896.

At a conference of the Association of Edison Illuminating Companies at Manhattan Beach, New York, in 1896, Ford sat next to Edison, who had long pioneered the idea of an electric automobile. When Ford finished outlining his gasoline-powered car, Edison banged his fist down on the table.

"Young man, that's the thing," he said. "You have it. The self-contained unit carrying its own fuel with it. Keep at it."[15]

Ford quit Detroit Edison two years later and set up his own company manufacturing automobiles. The two men became firm friends and often spent time in each other's houses. After Ford helped Edison out in 1915, when he had been burnt out, Ford planned to take Edison on a camping trip with tire manufacturer Harvey S. Firestone (1868 – 1938) and essayist John Burroughs (1837 – 1921). Though Ford was detained by business, he arranged for Ford trucks to carry them and their supplies. It became a regular event and by 1919, the convoy numbered fifty vehicles.

They were followed by herds of photographers and were visited at their campsite by President Warren Harding. Edison gave up attending in 1921, finding that there were just too many people involved.

Thomas Edison (left) and Henry Ford on Edison's 80th birthday in 1927.

LIGHT'S GOLDEN JUBILEE

To celebrate the fiftieth anniversary of Edison's incandescent light bulb, in 1929, Ford dedicated a museum and a turn-of-the-century village that he was building in Dearborn, Michigan, to his friend and mentor. The Edison Institute of Technology (today known as the Henry Ford Museum) now exhibits a test tube said to contain Edison's last breath and among the historic buildings in Greenfield Village is the Smith's Creek Grand Trunk Railroad Depot where the young Edison, his printing press and laboratory had been tossed out of the train. Edison's Menlo Park laboratory was also reconstructed there, along with his Fort Myers' laboratory which Ford obtained from Edison by building him another one.

On October 21, 1929, Edison and his wife arrived at Dearborn in Ford's private railroad car for the ceremonial opening of the museum and "Light's Golden Jubilee," also attended by President Herbert Hoover, physicist Marie Curie, and Orville Wright, the airplane pioneer.[16]

When he reached the site, Edison bent down and picked up a handful of the earth, which Ford had had shipped in from New Jersey by the truckload.

"Hmm, the same damn old New Jersey clay,"[17] said Edison.

THE LONELINESS OF GENIUS

Edison then visited the reconstructed laboratory where he had spent the years 1876 – 82. The *Detroit Free Press* reported:

> *As he walked to a chair and sat down, his companions in the party remained where they stood, apart from him a dozen feet. No word was spoken; it was as if by common consent the spectators instinctively felt awe here, in the presence of an old man upon whom the memories of eighty-two years were flooding back. He sat there, silent, his arms folded, an indescribably lonely figure, lonely in the loneliness of genius, of one who somehow had passed the others, who no longer has equals to share the world, his thoughts, his feelings. For five, perhaps ten, minutes, the scene was unmarred by a word or a gesture, except that now and then Edison looked about him and his eyes dimmed. Suddenly he cleared his throat and the spell was broken.*[18]

Ford then handed Edison an old mortar and pestle that he had used at Menlo Park. They had been found in pieces on the original site and Ford had glued them together again himself. He asked Edison what he thought of the reconstruction.

"Well, you've got this just about ninety-nine and one half percent perfect," said Edison.

"What is the matter with the other one-half percent?" asked Ford.

"Well," said Edison, "we never kept it as clean as this."[19]

After Edison left the lab, Ford had the chair in which he had been sitting nailed in place.

LET THERE BE LIGHT

That night, 500 personal guests of Edison and Ford attended the "history in the remaking" in person at a plush banquet held in a brand-new, grand ballroom in Dearborn, Michigan. As part of the celebration, Edison had agreed to recreate the lighting of the first successful incandescent light bulb fifty years earlier. The re-enactment was broadcast live on the radio. Listeners across the country, huddled around their radios and dimmed their lights.

They were to turn their lamps back on only after the great inventor, once again, achieved illumination.

"The lamp is now ready," said the announcer, "as it was a half century ago. Will it light? Will it burn? Edison touches the wire. Ladies and gentlemen – it lights. Light's Golden Jubilee has come to a triumphant climax."[20]

As the filament glowed brighter, Edison proclaimed: "Let there be light."[21]

At the dinner in his honor afterwards, a

tribute came by short-wave radio from Albert Einstein in Berlin. "The great creators of technics, among whom you are one of the most successful, have put mankind into a perfectly new situation, to which it has as yet not at all adapted itself,"[22] he said. Einstein was still living in Germany. Three years later, he too would become a resident of New Jersey.

Tributes also came from the Prince of Wales, Germany's President Paul von Hindenburg and the chairman of General Electric.

As he went to bed that night, he said: "I am tired of all the glory, I want to get back to work."[23]

Which is exactly what he did, working with Fred Ott on his home-grown rubber project. The first few samples of vulcanized rubber had been delivered when Edison fell into a coma. Four days later, in the early hours of October 18, 1931, the light in Edison's bedroom went on. The great man was dead.

TRIBUTES AND ACCOLADES

The flags on all Edison buildings flew at half-mast and the Pearl Street building that had housed the first power station carried a wreath. NBC paid homage with a musical tribute – as it had on his eightieth birthday.

Pope Pius XI expressed his deep sorrow and French Prime Minister Pierre Laval sent his condolences.[24] President Hoover said: "Mr. Edison was as great in his brave fight for his life as he was in the achievements which have made the whole world his debtor. I mourn his passing, not only as one of the greatest men our nation had produced, but as a personal friend."[25]

TESLA'S RESPECT

While Tesla was critical of Edison's methods, he paid tribute to the man himself.

> *The recurrence of a phenomenon like Edison is not very likely. The profound change of conditions and the ever increasing necessity of theoretical training would seem to make it impossible. He will occupy a unique and exalted position in the history of his native land, which might well be proud of his great genius and undying achievements in the interest of humanity.*

Edison's body lay in state for two days in the library of his West Orange laboratory where it was visited by dignitaries. Then on the evening of October 21, the lights were turned off all over America at 9.59 Eastern, 8.59 Central, 7.59 Mountain, and 6.59 Pacific. A minute later, with the flip of a switch, darkness in the US was banished forever.

Thomas Edison, Henry Ford, and Francis Jehl in the reconstructed Menlo Park Laboratory during Light's Golden Jubilee, 1929.

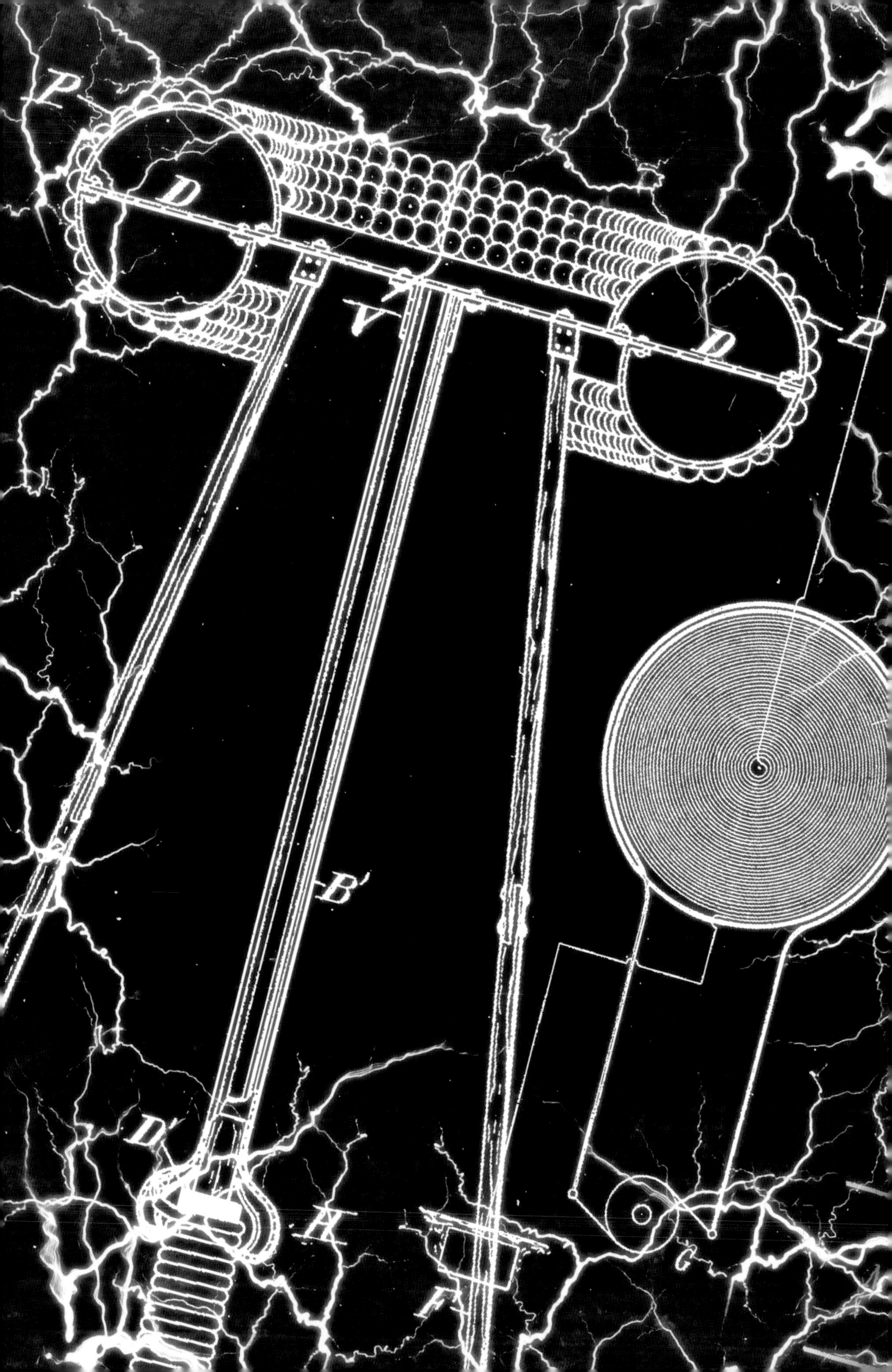
P
D
V
D
P
B'
D'
K
F
G

PART SIX

TESLA AGAINST THE WORLD

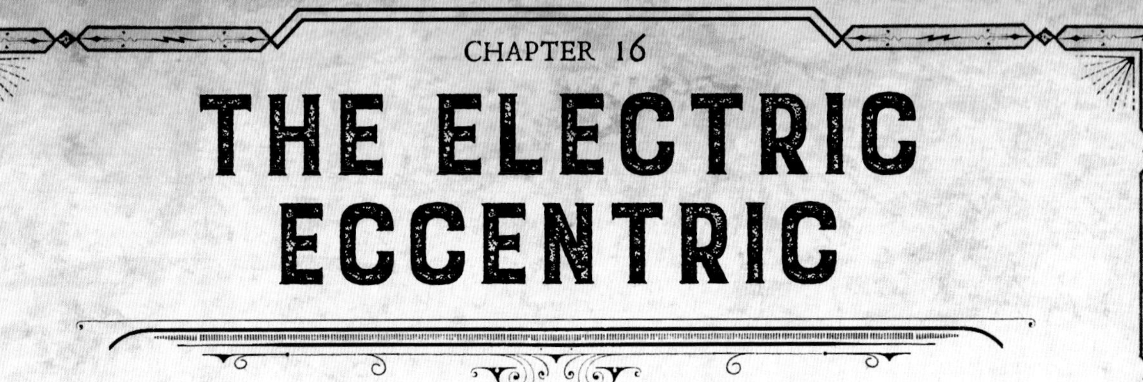

CHAPTER 16

THE ELECTRIC ECCENTRIC

While Edison was also an entrepreneur who knew how to bring his inventions to the marketplace, Tesla was a visionary and, without the help of a businessman like Westinghouse, his inventions became ever more weird and wonderful, and his pronouncements made him sound like a mad scientist.

He went to work developing Elisha Gray's teleautography, using a spinning disk with a spiral of holes in it to scan pictures. In 1902, this was used to send the first fax, but Tesla talked of sending pictures wirelessly at a time when wireless telegraphy had yet to be perfected. Nevertheless, Scottish inventor John Logie Baird (1888 – 1946) did use the spinning disk system when he developed his mechanical television system in 1926.

EXTREME EXPERIMENTS

When William Preece cancelled the test of Tesla's wireless equipment, Lloyds of London contacted Tesla and asked if he would rig up a ship-to-shore system for an international yacht race in 1896. Tesla refused, fearing that his work would be confused with the amateurish efforts of others in the field.[1] At the time, he could receive signals from his lab in East Houston Street fifty miles up the Hudson River at West Point.

He talked of using wind power, tidal power, solar energy, geothermal energy, and clean energy produced by recombining hydrogen and oxygen separated from water by electrolysis when no one was concerned with such things. He even suggested separating

Elisha Gray's teleautography machine, 1908.

nitrogen from the air by electrolysis, giving farmers abundant fertilizer.

Tesla lectured on X-rays at the New York Academy of Science and, when he was told that Marconi had successfully transmitted a radio signal eight miles, he talked of transmitting signals around the globe, through the Earth itself or through the ionized layers thought to exist in the upper atmosphere. However, he did not have any money to pursue these extreme ideas. Though his polyphase system was now powering the subway system, he was not making a cent from it.

However eccentric he seemed, Tesla was far from being a spent force. Attaching one of his oscillators to a support beam in his lab on Houston Street, he set off an artificial earthquake that brought the New York Fire Department running. He had already taken a sledgehammer to the machine, but Tesla told reporters that, by using the same method, he could split the Earth, destroying mankind.[2]

BEAUTIFUL UNFINISHED INVENTIONS

During the Spanish-American War of 1898, Tesla developed a remotely controlled boat that could be packed with explosives and rammed into enemy shipping. He proposed other "devil automata," even offering unmanned war machines to the Czar of Russia. Tesla said:

> *The continuous development in this direction must ultimately make war a mere contest of machines without men and without loss of life, a condition which would have been impossible without this new departure, and which, in my opinion, must be reached as preliminary to permanent peace.*[3]

In the *Electrical Engineer*, in an article headed "His Friends to Mr. Tesla," Thomas Martin urged Tesla to complete a long list of "beautiful but unfinished inventions," but he

THE REMOTE CONTROL BOAT

The remotely controlled craft was powered by large batteries on board. Radio signals activated switches, which controlled the boat's propeller, rudder, and running lights. Tesla had to invent a new kind of coherer or a radio-activated switch to achieve this. Each signal advanced a disc one step, making a new set of contacts that operated levers, gears, springs and motors. Tesla thought this system could be used on radio-guided torpedoes, but it was too far ahead of its time. The US Navy did finance some trials in 1916, but by then Tesla's patent had expired and, as the twentieth century progressed, many more uses would be found for remote control technology.

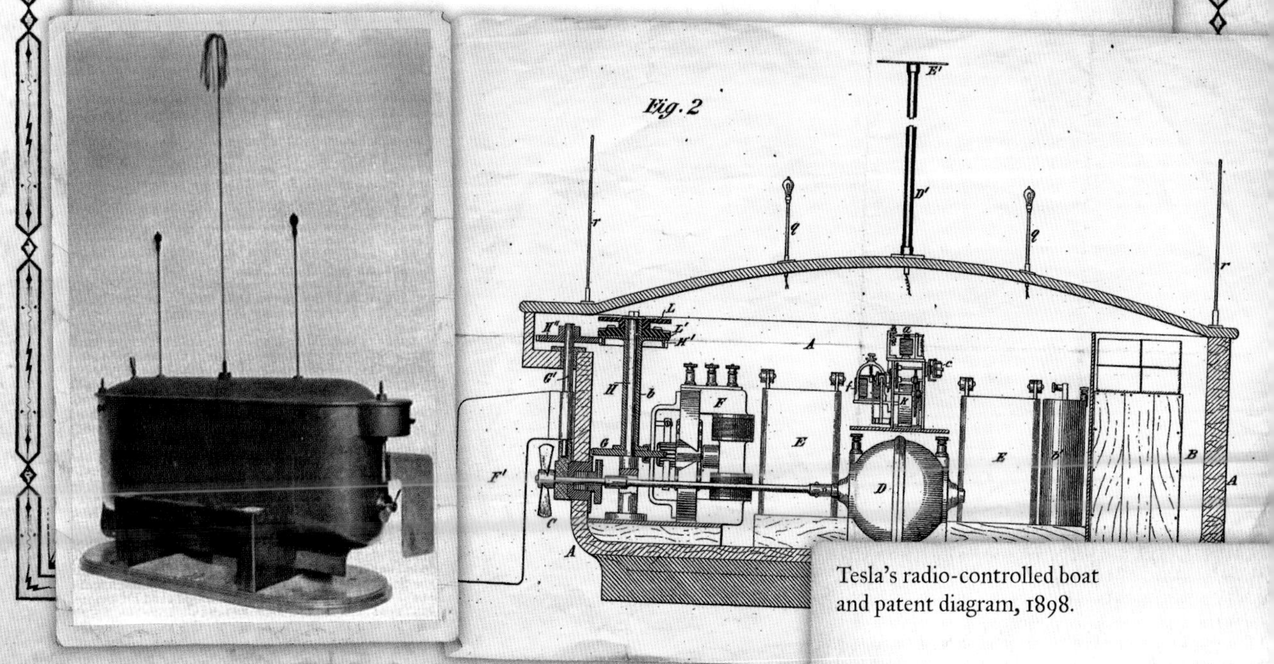

Tesla's radio-controlled boat and patent diagram, 1898.

should stop making statements about such fantastic things as remote-control aircraft that would "explode at will ... [and] never make a miss."[4]

Undaunted, Tesla claimed to have made a "dirigible wireless torpedo," a small remote-controlled airship.

But Martin had a point. Tesla's oscillator was not a commercial success. His fluorescent tubes never went on the market. His wireless transmission of power was never realized and he never got to build what he envisioned as "the first of a race of robots, mechanical men which will do the laborious work of the human race."[5]

However, not all Tesla's idea were mad. He boasted that he had produced a lamp that was far superior to the incandescent bulb, using one-third of the energy. He claimed:

> *As my lamps will last forever, the cost of maintenance will be minute. The cost of copper, which in the old system is a most important item, is in mine reduced to a mere trifle, for I can run on a wire sufficient for one incandescent lamp more than a thousand of my own lamps, giving fully five thousand times as much light.*[6]

On the strength of this, Tesla's friend John Jacob Astor invested $100,000 in the Tesla Electric Company and Tesla moved into the Waldorf-Astoria.

TESLA IN COLORADO

With Tesla's Coils now generating up to four million volts with sparks jumping from the walls to the ceilings, Tesla's Houston Street laboratory was becoming a fire hazard. Nor was it secure against the snooping of Edison's spies.

So, in May 1899, he moved out to Pike's Peak outside Colorado Springs where Westinghouse patent attorney, Leonard E. Curtis, arranged for him to get free power from the local utility, the El Paso Power Company.

Nikola Tesla in his Colorado Springs laboratory in December 1899, while his magnifying transmitter generates huge bolts of electricity.

In an empty field he built an experimental station, ringed with a fence and signs on it saying: "KEEP OUT, GREAT DANGER." Above the door was a phrase from Dante's *Inferno* said to be the inscription above the entrance to Hell: *"Abandon hope, all ye who enter here."* [7]

On top of his laboratory, Tesla built a 25-foot (8-meter) tower and a telescopic mast that raised a copper ball to a height of 142 feet (43 meters). Under it, he built a "magnifying transmitter", which was essentially a huge Tesla Coil. It was fed by a 50-kilowatt Westinghouse transformer that stepped the voltage up to 20,000 or 40,000 volts. With this, he experimented with sending electromagnetic signals through the rarefied air 5,000 feet (1,500 meters) above the Earth.

"There is nothing novel about telephoning without wires to a distance of five or six miles [eight or nine kilometers]," Tesla told Astor. "In this connection, I have obtained two patents." [8]

The magnifying transmitter could be cranked up to produce streams of artificial lightning 16 feet (5 meters) long and had set fire to the building more than once, but he found the danger exhilarating.

IS THERE LIFE ON MARS?

Tesla also experimented sending electrical signals through the earth and developed sensitive equipment to track the electrical storms that were common in late spring in that part of America. His equipment also picked up regular pulses, which he first thought were from aliens, possibly Martians. America was in the grip of Martian fever at the time.

The noted astronomer Percival Lowell (1855 – 1916) was studying the "canals" on Mars. In his book *Mars* (1895), Lowell concluded that there had been a drought on Mars and the Martians had built the canals to carry water from the polar ice caps. But when Tesla wrote of the possibility of communicating with aliens on other planets, it brought criticism.[9] However, Marconi also claimed to have detected radio signals coming from outside the atmosphere and tried to communicate with aliens.

COLD SHOULDERING THE POWERS-THAT-BE

While he was conducting his arcane experiments in Colorado, Tesla was contacted by the US Lighthouse Board. They wanted to install a wireless radio set on board the Nantucket Lightship so that it could give advanced warning of incoming shipping to New York and other east-coast ports. Tesla initially agreed to supply some experimental equipment, but was furious when he discovered that his apparatus was going to be tested against Marconi's.[10] His vision for inventions never did extend to grasping golden opportunities.

Tesla then insisted that he could not take a break from the important work he was doing in Colorado, unless the board put in an order for at least twelve wireless sets. They wouldn't and Tesla missed a great opportunity to demonstrate his equipment to the government. When the US Navy bought its first wireless equipment, it purchased it from German and French companies.[11]

GIANT SPARKS AND BLACK OUTS

As Tesla continued to increase the power of his magnifying transmitter, he became a danger to all around him. Arriving for work one day in mid-autumn, his assistant Kolman Czito found Tesla, the inventor, watering the ground around the metal plate he had buried near the lab as an earth. He gave Czito a pair of rubber soled shoes and put on a pair himself.

"All the way today, sir?" asked Czito.

"To the limit, my friend," said Tesla. "Now remember, keep one hand behind your back at all times."

This was to prevent a circuit being created

between his two arms that would send a lethal current through the heart.

Tesla then tottered out of the lab on his high shoes. Across the field he had placed his "cold lamps" that lit without any electrical connections and other test equipment. As he walked, sparks jumped from the ground. He positioned himself on a knoll about a mile from the lab where he could see the top of the mast.

It was evening and the lights were going on in Colorado Springs when Tesla gave the signal for Czito to close the switch. Streamers of lightning shrouded the mast high above the roof of the lab. A low rumble built to a roar of thunder that could be heard 15 miles (24 kilometers) away.

Suddenly there was a deafening silence. Below, Colorado Springs was plunged into darkness. Tesla phoned the power station.

Tesla's Experimental Station in Colorado Springs, 1899.

"You have cut off my power!" he yelled. "You must give me back my power immediately!"

The man on the other end of the phone explained that Tesla had short-circuited the generator. It was on fire. Fortunately, the powerhouse had a standby generator to turn the lights back on in Colorado Springs, but he refused to supply Tesla with any more electricity.[12] Soon after, Tesla left Colorado Springs claiming that his experiments had been a success. But, in reality, it was never clear what the results were.

TESLA MEETS MARCONI

Marconi was in New York when Tesla returned to the comfort of the Waldorf-Astoria in January 1900. The two men met up.

"I remember him when he was coming to me asking me to explain the function of my transformer for transmission of power to great distances," said Tesla. "Mr. Marconi said, after all my explanations of the application of my principle, that it is impossible."

"Time will tell, Mr. Marconi," Tesla replied.[13]

Tesla went to Washington in the hope that the US Navy or Coast Guard might buy his wireless transmitters, and planned to prove his system by transmitting a signal across the Atlantic.[14] Westinghouse, though now in financial difficulties, fronted the money. He sent an agent to Britain to find a suitable site for a receiving station. In 1900, Tesla filed three patents on wireless communication and fine-tuned his plans for a trans-oceanic broadcasting system.

J.P. MORGAN MAKES AN OFFER

Tesla moved in high circles. In the autumn of 1900, he was invited to attend the wedding of Louisa, the daughter of Wall Street magnet J.P. Morgan, who was a keen yachtsman and commodore of the New York Yacht

Club. During the America's Cup, he offered Marconi $200,000 for his American patents, including the "ocean rights ... if ever wireless telegraphy could communicate from England to New York." But the deal fell through and Morgan opened negotiations with Tesla.

Morgan agreed to give Tesla $150,000 to build a transatlantic transmitter with a 90-foot (27 meter) tower, in return for fifty-one percent of the company and the patents. The lighting patents that Astor had an interest in were added to this later.[15]

WORLD TELEGRAPHY CENTER

To build his transatlantic transmitter, Tesla bought Wardenclyffe, a tract of land at Shoreham on Long Island Sound. He was planning to build a 1,800-acre (728-hectare) "Radio City" which Stanford White set about designing. Traveling out to the site by train one morning, Tesla read an article in *Electrical Review* in which Marconi admitted to using a Tesla Coil in his wireless experiments.[16]

Furious, Tesla immediately trashed his plans to build a modest 90-foot tower and started designing a 600-foot (183-meter) edifice. These grandiose plans did not go down well with Morgan and White scaled them down.[17]

MARCONI'S MIRACLE

Marconi had already installed a power transmitter with a 200-foot (60-metre) mast at Poldhu in Cornwall, England, and had sent test transmissions to Crookhaven in Ireland, 200 miles (320 kilometers) away. Meanwhile, the sister station was being built on Cape Cod. Both were flattened by storms in September 1901.

The Cornish station was rebuilt, but the aerial on the other side of the Atlantic was to be strung from a kite flown from Signal Hill in Newfoundland. On December 12, 1901, it picked up a signal – three dots, the Morse code for the letter S, from Poldhu, proving that transatlantic wireless communication was possible.

Otis Pond, an engineer then working for Tesla, said, "Looks as if Marconi got the jump on you." Tesla replied, "Marconi is a good fellow. Let him continue. He is using seventeen of my patents."[19]

Thomas Martin held a banquet for three-hundred guests to celebrate Marconi's achievement at the Waldorf-Astoria. Tesla did not attend and ducked out of the hotel before Marconi arrived. However he did send a letter saying "he could not rise to the occasion." Nevertheless he congratulated Marconi on his brilliant results.[20]

In his speech, Marconi pointed out that his wireless was already installed on over seventy ships – thirty-seven in the British Royal Navy, twelve in the Italian Navy and the rest on liners belonging to Cunard, North German Lloyd, and others. There were already twenty stations in operation on land in Great Britain and more in construction. And he acknowledged his indebtedness to "Clerk Maxwell, Lord Kelvin, Professor Henry and Professor Hertz."[21] But not Tesla, whose claims became ever more grandiose – he was now promising that the press would be tied into his wireless system,

TESLA'S CELLPHONE

As long ago as 1900, Tesla wrote of a world system of wireless transmission:

> *The World-System has resulted from a combination of several original discoveries made by the inventor in the course of long continued research and experimentation. It makes possible not only the instantaneous and precise wireless transmission of any kind of signals, messages or characters, to all parts of the world, but also the inter-connection of the existing telegraph, telephone, and other signal stations without any change in their present equipment. By its means, for instance, a telephone subscriber here may call up and talk to any other subscriber on the globe.*[18]

This is surely the cellphone network we have over a century later.

and every customer would be able to print their own newspaper.[22] Back then it sounded crazy, but it is not far removed from what we can do today, downloading and printing from home computers.

WARDENCLYFFE

The ideal experimental station Tesla really wanted to build would have cost $1 million. Nevertheless what he ended up building with the $150,000 Morgan had given him was impressive. The tower rose 187 feet (57 meters) in the air. On the top was a 57-ton steel sphere. Under the tower was a shaft that plunged 120 feet (36 meters) into the ground. Sixteen iron pipes were driven down another 300 feet (91 meters).

"It is necessary for the machine to get a grip of the Earth, otherwise it cannot shake the Earth," Tesla explained. "It has to have a grip ... so that the whole of this globe can quiver."[23]

But construction was already running into difficulties when Marconi sent his transatlantic signal. While he was using Tesla's patents, his equipment was cheap by comparison. The stock market crashed and Morgan cut off the money.

Nevertheless, at the end of July 1903, Tesla finally switched on his magnifying transformer. Once the mushroom-shaped cupola became fully charged, local villagers heard a rumble of thunder and a strange light appeared above Tesla's tower. It could even be seen on the shores of Connecticut, the other side of Long Island Sound.

But the station was not in operation long. Soon Westinghouse creditors came to cart away the heavy equipment and Tesla's tower again fell silent. Tesla had made the fundamental mistake of promising to be able to transmit unlimited amounts of power for free and the captains of industry and the moguls of Wall Street were not having that.

TESLA'S WORLD-WIDE BRAIN

Tesla responded with an article published simultaneously in *Electrical World* and *Scientific American* in which he perfectly described today's internet:

> *The results attained by me have made my scheme of intelligence transmission, for which the name of "World Telegraphy" has been suggested, easily realizable. It constitutes a radical and fruitful departure from what has been done heretofore ... It involves the employment of a number of plants, all of which are capable of transmitting individualized signals to the uttermost confines of the earth. Each of them will be preferably located near some important center of civilization and the news it receives through any channel will be flashed to all points of the globe. A cheap and simple device, which might be carried in one's pocket, may then be set up*

Tesla's Tower at Wardenclyffe, Shoreham, Long Island.

somewhere on sea or land, and it will record the world's news or such special messages as may be intended for it. Thus the entire earth will be converted into a huge brain, as it were, capable of response in every one of its parts. Since a single plant of but one hundred horsepower can operate hundreds of millions of instruments, the system will have a virtually infinite working capacity, and it would immensely facilitate and cheapen the transmission of intelligence. The first of these central plants would have been already completed had it not been for unforeseen delays ...[24]

BEING A WIRELESS FOOTNOTE

While Tesla was working on his great vision, more modest men were moving into the field. In 1901 Lee De Forest (1873 – 1961), who had once applied to be Tesla's assistant, sent wireless messages across the Hudson River. He speeded up the rate of transmission to thirty words a minute – that was as fast as any Morse-code operator could send them. In 1904, he sent a signal from Buffalo to Cleveland, a distance of 180 miles (290 kilometers). Then in 1908, he transmitted across the Atlantic.

Marconi and De Forest were still sending messages in Morse code, but Canadian-born Reginald Fessenden (1866 – 1932) realized that it was possible to modulate a radio signal and, in 1906, he transmitted music down the Massachusetts coast. In 1910, De Forest broadcast the voice of Italian opera singer Enrico Caruso (1873 – 1921) from the Metropolitan Opera House in New York. This marked the beginning of radio as a medium of entertainment, but Tesla took him to court for patent infringement and won.

Nevertheless progress continued apace. By 1909, Lee De Forest had perfected the radio-telephone which had been adopted by the navies of the US, Britain, and Italy. Tesla was rapidly becoming a footnote in the history of wireless.

THE MILLION-DOLLAR FOLLY

The newspapers began to call Wardenclyffe "Tesla's million-dollar folly." Tesla had a nervous breakdown and retreated to his room at the Waldorf-Astoria where he nursed an injured pigeon he had found near the New York Public Library. However, at night, he sometimes stole out to Wardenclyffe to hook himself up to the high-frequency machinery.

"I have passed 150,000 volts through my head," he told *The New York Times*, "and did not lose consciousness, but I invariably fell into a lethargic sleep sometime after."[25]

100 YEARS AHEAD OF HIS TIME

In "The Future of the Wireless Art" published in *Wireless Telegraphy and Telephony* in 1908, Tesla described his vision of the future:

> *As soon as it is completed, it will be possible for a business man in New York to dictate instructions, and have them instantly appear in type at his office in London or elsewhere. He will be able to call up, from his desk, and talk to any telephone subscriber on the globe, without any change whatever in the existing equipment. An inexpensive instrument, not bigger than a watch, will enable its bearer to hear anywhere, on sea or land, music or song, the speech of a political leader, the address of an eminent man of science, or the sermon of an eloquent clergyman, delivered in some other place, however distant. In the same manner any picture, character, drawing, or print can be transferred from one to another place. Millions of such instruments can be operated from but one plant of this kind. More important than all of this, however, will be the transmission of power, without wires, which will be shown on a scale large enough to carry conviction.*[26]

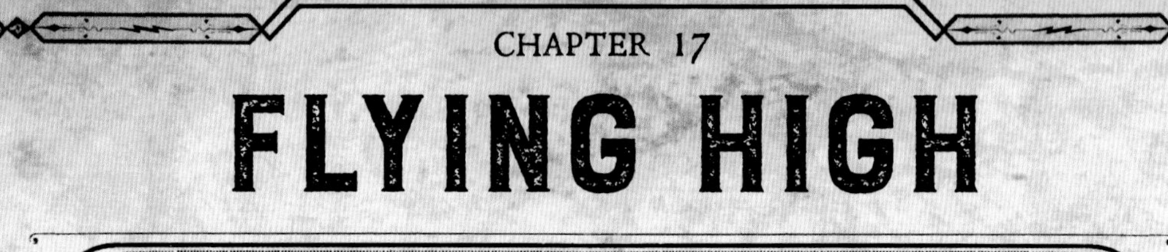

CHAPTER 17

FLYING HIGH

The Wright brothers had made their first powered flight in 1903. Astor was keen on flying machines and encouraged Tesla to take an interest. At a dinner at the Waldorf-Astoria, at the beginning of 1908, Tesla made another of his pronouncements of the future: "Aerial vessels of war will be used to the exclusion of ships."[1] He went on to say "the propeller is doomed." It would, he said, have to be replaced by "a reactive jet."

That June, he revealed that he was working on a heavier-than-air machine of his own. Three years later he spelt out his plans:

> *My flying machine will have neither wings nor propellers. You might see it on the ground and you would never guess that it was a flying machine. Yet it will be able to move at will through the air in any direction with perfect safety, higher speeds than have yet been reached, regardless of weather and oblivious of holes in the air or downward currents. It will ascend in such currents if desired. It can remain absolutely stationary in the air, even in a wind, for great length of time. Its lifting power will not depend upon any such delicate devices as the bird has to employ, but upon positive mechanical action.*[2]

Naturally his plane would be powered by electricity, transmitted wirelessly from ground stations – Tesla was developing not so much a helicopter, but a vertical take-off and landing vehicle (VTOL).

TAKING-OFF VERTICALLY

Having started work on it in 1908, the patents for his brilliantly designed flying machine would not be filed until 1921 and 1927. Dubbed the "flivver plane" – flivver being early twentieth-century slang for a cheap car – it was said to combine the qualities of a helicopter and a plane, and could fly vertically as well as horizontally. According to a press report: "It is a tiny combination plane which, its inventor asserts, will rise and descend

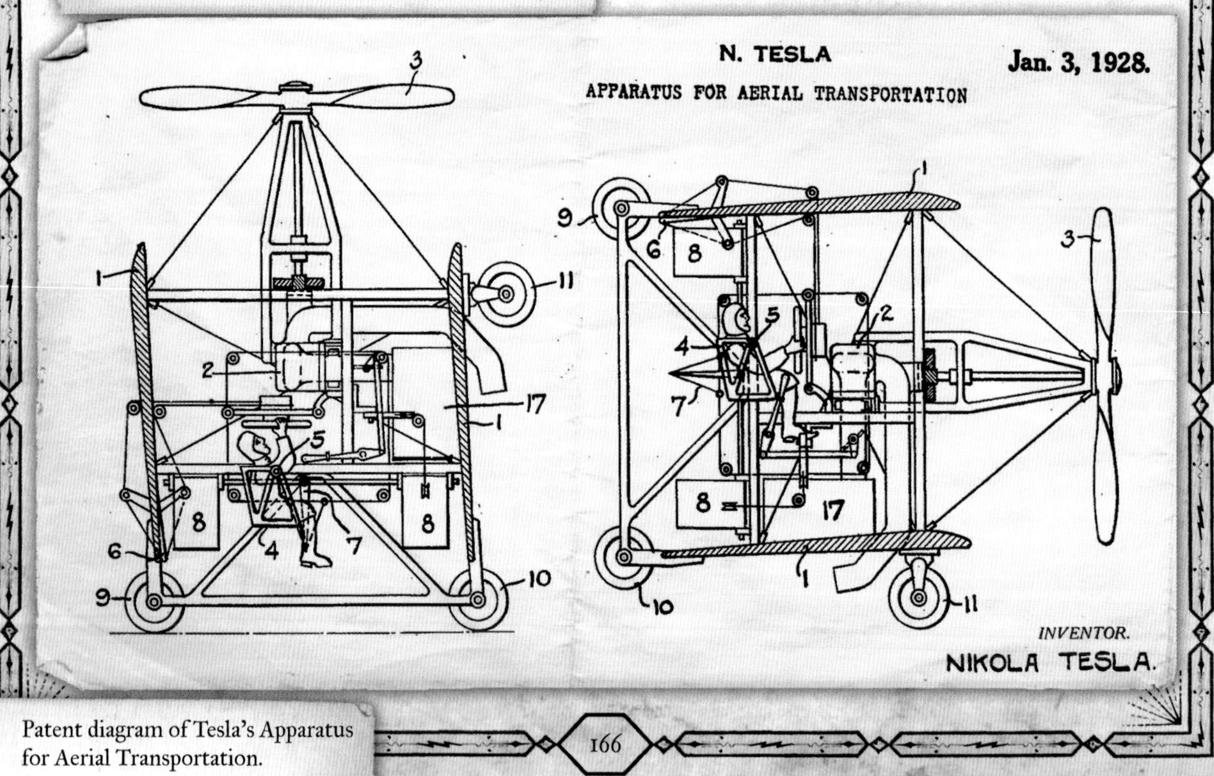

Patent diagram of Tesla's Apparatus for Aerial Transportation.

vertically and fly horizontally at great speed, much faster than the speed of the planes of today. But despite the feats which he credits to his invention, Tesla says that it will sell for something less than $1,000."[3]

It would carry a pilot and three or four passengers. Flying horizontally like a plane, while landing vertically like a helicopter. By the time patents were finally granted in 1928, Tesla was seventy-two. But no full-scale model was ever built.

Tesla's vertical take-off concept gathered dust until the 1950s when Lockheed and Convair tested vehicles that, although vastly more advanced in engineering, stuck to Tesla's fundamental plans. Although tests were relatively successful, no planes went into production. Technology had still not caught up with Tesla. But the potential military and commercial advantages of VTOL were too great to be ignored. By the 1990s, leading vertical take-off fighter aircraft such as the Anglo-American Harrier and the Russian Yak-36 were fully operational.

That Tesla should imagine the VTOL flivver idea decades ahead of the jet engine, and with aviation in its infancy, is truly amazing. But after his death, many other Tesla drawings were found not only of aircraft, but also plans for jet propelled flying cars and sketches of interplanetary spaceships.

PERFECTING BLADELESS TURBINES

In 1909, Tesla invented the "bladeless turbine" which, he believed, would replace the gasoline engine in an automobile or would be used to power aircraft, ships or torpedoes.

In his breakthrough induction motor, Tesla's rotating magnetic field dragged the rotor around. Similarly Tesla believed that it would be possible to use a moving stream of steam or compressed air to turn a series of discs attached to a shaft. The fluid would enter at the edge of the disc and exit at the central shaft. As the fluid spiral down between the discs, it would drag them around with it. If the action was reversed, the turbine acted as a pump with the fluid spiraling from the center outward.

Without blades, the engine would be cheaper to build and easier to maintain, and

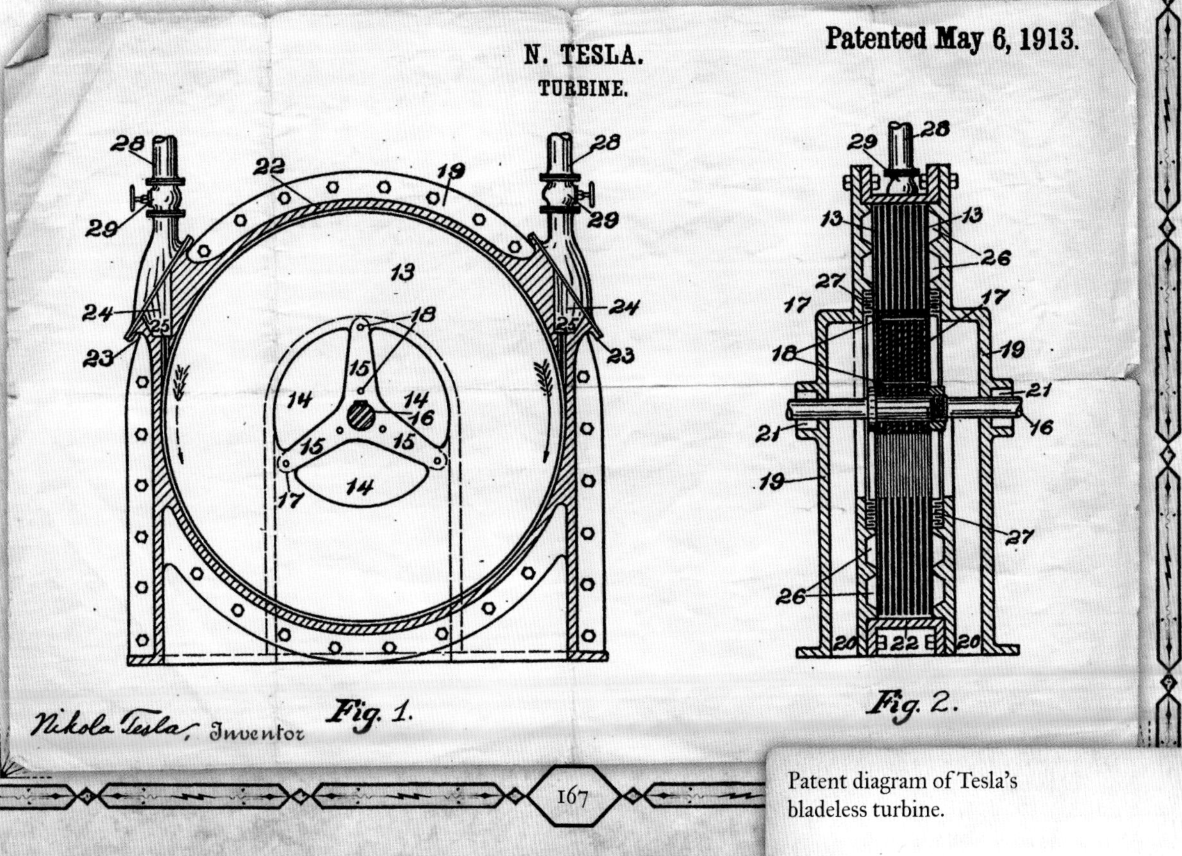

Patent diagram of Tesla's bladeless turbine.

gave a vastly improved power to weight ratio, making it perfect for flying.

"I have accomplished what mechanical engineers have been dreaming about ever since the invention of steam power," said Tesla. "That is, the perfect rotary engine."[4]

Tesla was confident that his new engine would be so successful that he would have the money to reopen Wardenclyffe.

TESLA'S PROPULSION STALEMATE

The first prototype had eight discs each 6 inches (15 centimeters) in diameter. Weighing less than 10 pounds (4.5 kilograms) it developed 30 horsepower and would rotate at up to 35,000 revolutions per minute. In 1909, Tesla filed two patents – one for the turbine, the other for the pump.

Tesla set up the Tesla Propulsion Company with Joseph Hoadley, whose Alabama Consolidated Coal and Iron company planned to use the Tesla pump as a blower in its blast furnaces. To promote his new invention, Tesla arranged to give a public demonstration at the Waterside Station of the New York Edison Company. Tesla had two engines built with 18-inch (46-centimeter) rotors. They were just 3 feet (90 centimeters) long, 2 feet (60 centimeters) wide and 2 feet (60 centimeters) high, and weighed 400 pounds (181 kilograms). Revolving at 9,000 rpm, they each developed 200 horsepower.

In the demonstration, the two motors would engage in a tug-of-war, with the power they developed registering on a torque rod connecting them. But that meant the motors did not actually turn, just strained against each other. The audience were not impressed and the story soon circulated that the test was a complete failure.[5]

OZONE THERAPY

It had long been thought that ozone had healing properties and Tesla patented an ozone-generating machine in 1896, forming the Tesla Ozone Company in 1900. His portable "ozonizers" were used to disinfect rooms as ozone kills germs. He is also thought to have ozonated olive oil. He gave one of his machines to his long-serving secretary George Scherff for his wife who was ill at the time. "I believe that it will do you and Mrs. Scherff a lot of good," he wrote, "unless you have no electricity supply circuit in your house, in which case, it will be necessary to move into other quarters."

VANISHING INVESTORS

Scientifically, the test had been a success, but Tesla needed money to develop it. Westinghouse was out of business. In 1912, Astor drowned on the *Titanic*. The following year J.P. Morgan also died. After his funeral, Tesla approached his son Jack, who loaned the inventor $20,000. Tesla then tried to sell his turbine to Sigmund Bergmann (1851 – 1927), an old colleague of Edison's who had set up a large manufacturing concern in Germany. But the deal was not concluded before the start of World War I. Then Jack Morgan became involved in helping Britain and France finance the war and lost interest.[6]

REMOTE CONTROLLED MECHANICAL DOG

Tesla was bitter when Marconi won the Nobel Prize in 1909, for radio communications, something that Tesla considered to be his invention. He then discovered that, while working with Alexander Graham Bell and Tesla's former assistant Fritz Lowenstein (1874 – 1922), the inventor Jack Hammond (1888 – 1965) had built an electric dog that worked by remote control.[7]

While Hammond assured Tesla that he had not infringed any of his patents, Tesla insisted that he get a share in any profits. The two of them formed the Tesla-Hammond Wireless Company funded by Hammond's father, mining magnate John Hays Hammond (1855

– 1936). Hammond was interested in Tesla's patents on selective tuning, which divided up the radio spectrum into numerous channels – a crucial development in radio.[8] He contacted the War Department with the proposal for a ship-to-shore communication system and hired Fritz Lowenstein and electrical engineer Benjamin Franklin Miessner (1890 – 1976) to form a military research group at the family estate in Gloucester, Massachusetts. Tesla was sidelined again.[9]

OUTRAGEOUS CLAIMS

At the convention of the National Electric Light Association in 1911, Tesla said that he would be able to run the streetcars in Dublin from a power station in Long Island City. His wireless transmitter would generate enough power to light the entire United States.[10]

Tesla also claimed to have perfected a new steam engine, a tiny turbine that weigh only one pound and produced ten horsepower.[11] He made overtures to the Japanese Imperial Navy, in the hope that they would take his turbines to power their torpedoes. He also had meetings with GE and the Seiberling Company, who developed power boats. Ford were approached with prototype car engines, while Kaiser Wilhelm II showed some interest in their military applications. But Hammond refused to come up with any more development money.

Instead Hammond made a deal with the Radio Corporation of America – later RCA – shortly after Tesla's wireless patents had run out. Again the inventor had failed to profit from his inventions.

The Electric Dog with light-controlled remote guidance, 1912.

But Tesla was never idle for long. An experiment with his Tesla Coils in Sweden had demonstrated that children in an electrified environment grew more quickly and scored higher in aptitude tests. So Tesla went to work for the superintendent of New York's public schools installing Tesla Coils in the walls of a school for a pilot study.[12]

SURROUNDED BY LAWYERS

When the *Titanic* sank, Marconi was credited with saving the lives of the 710 survivors, as it was his equipment that had summoned the rescue ships. This was so galling for Tesla that he began suing Marconi for patent infringement.

While Telefunken in Germany had also infringed Tesla's patents, it was too powerful to sue. However, when the company came to America to set up transatlantic stations at Tuckerton, New Jersey, and Sayville, New York, it sought out Tesla in the hope that they could present a united front to Marconi. Then, in 1914, Tesla was approached by the American Marconi company. But they only offered stock; Tesla needed cash. He appealed to Jack Morgan for help, saying the US government had already installed $10-million-worth of his equipment and he was expecting to receive compensation.[13]

Meanwhile Telefunken was suing Marconi who, in turn, was suing Lowenstein and the US Navy. However, as the wireless equipment Hammond supplied the War Department was being used to test guided missiles, it was classified. So Hammond was immune from litigation.

WORLD WAR I

With the outbreak of World War I in 1914, the British cut Germany's transatlantic cables. Consequently, the Telefunken stations in America became of vital importance. Fearing that they may be used to direct the movements of battleships and submarines, the British wanted them shut down. While ostensibly neutral, President Woodrow Wilson (1856 – 1924) signed a bill taking them over. Because of his connections with Telefunken, Tesla also came under suspicion.

As the patent battles continued, Marconi arrived in New York onboard the *Lusitania*, which was sunk by the Germans on the return journey.[14] As Italy was on the Allied side in World War I, public opinion was on Marconi's side in the dispute. Nevertheless, Tesla had his day in court but, because of the war, the legal battle was abandoned before the case was decided.[15]

When the US joined World War I in 1917, Tesla's monthly royalty checks from Telefunken stopped. Edison, Fessenden, and others got jobs as advisors to the government, but Tesla, always the outsider, was left out.[16]

THE NOBEL PRIZE

On November 6, 1915, *The New York Times* mistakenly announced that Tesla and Edison had been jointly awarded the Nobel Prize for physics. Although he had received no official notification, Tesla foolishly told the *Times* that he thought he had been awarded the Nobel Prize for a device he had filed a patent for a month earlier that made it "practicable to project the human voice not only for a distance of five thousand miles, but clear across the globe."[17] He then explained how it would work:

> *The plant would simply be connected to the telephone exchange of New York City and a subscriber will be able to talk to any other telephone subscriber in the world, and all this without any change in his apparatus. This plan has been called my 'world system'. By the same means, I propose also to transmit pictures and project images, so that the subscriber will not only hear the voice, but see the person to who he is talking.*[18]

Tesla agreed that Edison deserved a dozen

Nobel Prizes, though he said he had no idea which one Edison had been awarded the prize for.

The story turned out to be just a rumor. It was the Nobel Prize that never was. The reality was that Tesla had not even been nominated, though Edison had been. Tesla was not nominated again until 1937 and did not get it then either.

THE BOLTS OF THOR

On December 8, 1915, in an article "Tesla's New Device Like Bolts of Thor," *The New York Times* reported that Tesla was taking out a patent on a "manless airship" that had neither an engine nor wings and could be sent at a speed of 300 miles (480 km) a second to any place on the globe using electricity:

> *Ten miles or a thousand miles, it will all be the same to the machine, the inventor says. Straight to the point, on land or on sea, it will be able to go with precision, delivering a blow that will paralyze or kill, as is desired. A man in a tower on Long Island could shield New York against ships or army by working levers.*[19]

TESLA'S RADAR CONCEPT

While Tesla's ideas on unmanned airships and Bolts of Thor seem unworldly, he also described a way of detecting ships at sea. His idea was to transmit high-frequency radio waves that would reflect off the hulls of vessels and appear on a fluorescent screen. This was one of the first descriptions of what we now call radar, but was too far ahead of its time to be taken seriously. However, when the French engineer Émile Girardeau (1882 – 1970) built an obstacle-locating radio apparatus in 1934, he said it was "conceived according to the principles stated by Tesla."[20]

Tesla may not have won the Nobel Prize, but in 1917 he grudgingly accepted the Edison medal. At the award ceremony, Tesla

TESLA<DOT>COM

In his autobiography, *My Inventions*, published in 1919, Tesla envisaged that in nine months, without undue expense, he could deliver:

- The interconnection of existing telegraph exchanges or offices all over the world;
- The establishment of a secret and non-interferable government telegraph service;
- The interconnection of all present telephone exchanges or offices around the globe;
- The universal distribution of general news by telegraph or telephone, in conjunction with the press;
- The establishment of such a "World System" of intelligence transmission for exclusive private use;
- The interconnection and operation of all stock tickers of the world;
- The establishment of a world system of musical distribution, etc.;
- The universal registration of time by cheap clocks indicating the hour with astronomical precision and requiring no attention whatever;
- The world transmission of typed or handwritten characters, letters, checks, etc.;
- The establishment of a universal marine service enabling the navigators of all ships to steer perfectly without compass, to determine the exact location, hour and speed; to prevent collisions and disasters, etc.;
- The inauguration of a system of world printing on land and sea;
- The world reproduction of photographic pictures and all kinds of drawings or records.[21]

Here we have the internet, GPS and Sat-nav, all conceived by Nikola Tesla, decades ahead of the first computer.

was compared to Faraday, but the medal was really being given to him for work he had done thirty years before.[22]

TRASHING THE TOWER

Tesla's wireless dreams died when he had to sign over Wardenclyffe to the Waldorf-Astoria as he could not pay his hotel bill which had reached $20,000 overdue ($400,000 today). In July 1917, Tesla left the Waldorf-Astoria where he had lived for twenty years. He moved to Chicago to continue his work on turbines at Pyle National.

The following month, the tower at Wardenclyffe was blown up. He had hoped it could have been used to locate and destroy enemy submarines. Instead, it was decided that it was too dangerous as it could be used for communication by enemy spies.[23]

With any further litigation on wireless patents on hold, American Marconi, AT&T, Westinghouse, and GE got together behind closed doors in Washington and formed RCA. At the end of the war, the Westinghouse Company set up independently using Marconi patents. In 1920, Tesla wrote, offering his services. They were refused. However, a little later, Westinghouse wrote again, asking Tesla if he would like to broadcast to their "invisible audience" one Thursday evening.[24]

Tesla replied dismissively that, twenty years earlier, he had promised J.P. Morgan that his "world system" would enable the voice of a telephone subscriber to be transmitted to any point on the globe. "I prefer to wait until my project is completed before addressing an invisible audience," he said.[25]

FLASHBACKS FROM THE FUTURE

Though Tesla's Wardenclyffe Tower was in ruins, the dream lived on in the artwork of Frank R. Paul. A long-time admirer of Tesla and editor of *Electrical Experimenter*, Hugo Gernsback, (1884 – 1967) hired Paul to show the world what Tesla's tower would have looked liked if it had been completed.

Paul's cover illustration added transmitters and Tesla's wingless flying machines zapping nearby ships with their death-rays. Tesla was so thrilled, he used the illustration as his letterhead.

The demolition of Wardenclyffe, 1917.

In 1919, *Electrical Experimenter* serialized Tesla's autobiography *My Inventions*. This too, was illustrated by Frank Paul's drawing, along with photographs of the equipment. This boosted the circulation of the magazine to around 100,000 and provided Tesla with a modest income. Earning nothing for his wireless patents, his only source of income was the Waltham Watch Company who were manufacturing the Air-Friction Speedometer he had designed.

Tesla's designs made a further move into science fiction when the 1931 horror classic *Frankenstein* used Tesla Coils to make lightning flashes. Much of the equipment used by Dr. Frankenstein bears an uncanny resemblance to the apparatus Tesla invented for his experiments. Indeed, Tesla favored the movie's producer Carl Laemmle as he fought patent battles with Edison when establishing Universal Pictures.[26]

MADE IN MILWAUKEE

Tesla was still hoping his bladeless turbines would make him rich. The US Machine Manufacturing Company asked about putting one in an airplane. The Chicago Pneumatic Tool Company also made enquiries.

The Allis-Chalmers Manufacturing Company of Milwaukee built three of Tesla's turbines. They made reciprocating engines, turbines and other heavy machinery. However, Tesla displayed his usual lack of tact and diplomacy and ruined the project from the outset. Insistent on interfering in negotiations with the senior staff, he went directly to the president of the company and, while engineers were preparing a feasibility report, he contacted the board of directors and sold them the idea before the engineers had had their say.

Three of Tesla's turbines were built. But vibrations caused them to crack. Tesla walked out. At the time, metallurgy was in its infancy. However, manufacturers have since made pumps using Tesla's principles, and his turbine works perfectly using disks made from advanced materials such as carbon fiber, titanium-impregnated plastic and Kevlar.

After a sojourn at Waltham Watches in Boston, he worked on a petrol-powered turbine at Budd Manufacturing in Philadelphia. And he was not without his other successes. He sold a motor that was used in cinema equipment to Wisconsin Electric and a "fluid diode" to an oil company that was said to be "the only valving patent

Cover of Hugo Gernsback's science magazine *The Experimenter*, March 1925, showing Tesla-style electrotherapy.

without moving parts."[27] Money began coming in, but never in the amounts that he over-optimistically predicted.

CRACKS START TO APPEAR

Returning to New York, Tesla moved into the Hotel St. Regis. He became more and more eccentric. He would circle the block three times before entering the hotel, avoiding stepping on the cracks in the sidewalk. And he was fanatical about cleanliness, except when it came to pigeons which he still fed outside New York Public Library.

Tesla claimed that he usually slept for just two hours a night, three being too much. It seems he had to go one better than Edison who claimed only to sleep four hours a night, though when he sat in his lab he would take two three-hour naps a day. Tesla probably did the same. Hotel staff said that they often found him sitting transfixed and they found they could work around him in his room without disturbing him.[28]

WARDENCLYFFE REVISITED

Tesla sued the Waldorf-Astoria over the destruction of Wardenclyffe. During the trial, Tesla lovingly described every detail of his experimental station. Though $350,000's worth of equipment had been destroyed to cover a debt of $20,000, the judge found in favor of the hotel.

While Italy's Fascist dictator Benito Mussolini (1883 – 1945) saluted Marconi, Vladimir Ilich Lenin (1870 – 1924), who had led the Bolshevik Revolution in Russia in 1917, made overtures to Tesla, asking him to come to the Soviet Union to build power stations and an AC distribution system.

Tesla was also invited to speak at a meeting of the Friends of Soviet Russia in Springfield, Massachusetts, traveling there with Ivan Mashevkief from the Russian Workers Club of Manhattan. According to an FBI agent at the event, Tesla "prophesied that Italy would soon adopt a communist form of government."[29] However, there is

Advertisement for the Waltham air-friction speedometer invented by Tesla.

no evidence that he knew what he was getting himself into. Tesla took little interest in politics and he was, at best, naïve.

BEING MOVED ON

Neglecting to pay for his room at the Hotel St. Regis, he moved to the Hotel Marguery on Park Avenue and 48th Street. Fortunately this was close to his night-time haunt, Bryant Park behind New York Public Library where he fed the pigeons.

In 1926, he moved on to the Hotel Pennsylvania. In his suite, he kept a pigeon he had found with a broken leg and wing.

"Using my mechanical knowledge, I invented a device by which I supported its body in comfort in order to let the bones heal,"[30] he said. He reckoned it cost him more than $2,000 to heal her. She was "one of the finest and prettiest birds I have ever seen,"[31] he said.

OVERTAKEN BY EVENTS

With development of his bladeless turbine reaching a dead end, Tesla returned to the idea of powering planes and cars remotely from large central power stations like the one he had tried to build at Wardenclyffe.

With the help of John B. Flowers, an inspector at an aircraft factory, he drafted a ten-page proposal that was sent to J. H. Dillinger, head of the Radio Laboratory at the Bureau of Standards in Washington, D.C. Flowers told Dillinger that Tesla's system would power a plane at any point around the world, and that Tesla had already developed an oscillator to provide the power. Tesla was willing to handover the secret to the American government if they agreed to build the plant.

A meeting in Washington was arranged and Dillinger sent the proposal to physicist Harvey L. Curtis (1875 – 1956), but Curtis simply picked holes in the basic science. Tesla contested it, but found that he was not up with the latest techniques in physics. Like Edison, Bell, and the Wright Brothers, time had caught up on Tesla. Einstein, with his Nobel Prize, was the new star in the scientific firmament.

POWERED BY COSMIC RAYS

Despite the rejection of their plans in Washington, Flowers and Tesla went to Detroit to try and sell Tesla's "Aeromobile" – his flying car – to General Motors. Tesla also tried to sell his speedometer to Ford at the same time, but its high cost made it best suited to luxury cars.

Tesla held talks with US Steel concerning installing his bladeless turbines on the exhaust from the blast furnaces, generating huge amounts of electricity. But, apparently, a test did not go ahead.

Then in Buffalo, Tesla conducted some top-secret experiments. It was said that the petrol engine in a Pierce-Arrow sedan was replaced by an AC induction motor. A "power receiver" using twelve valves was set in the dashboard connected to a six-foot (two-meter) antenna. There is other speculation that it was powered by a steam or petrol-driven turbine, but no physical evidence of either design has been found.

While experimenters were using Tesla Coils to try to split the atom, Tesla himself was making more outlandish predictions, saying that all the machinery on Earth could be powered by cosmic rays, which he claimed to have discovered while investigating X-rays and radioactivity in Colorado Springs in 1899.[32]

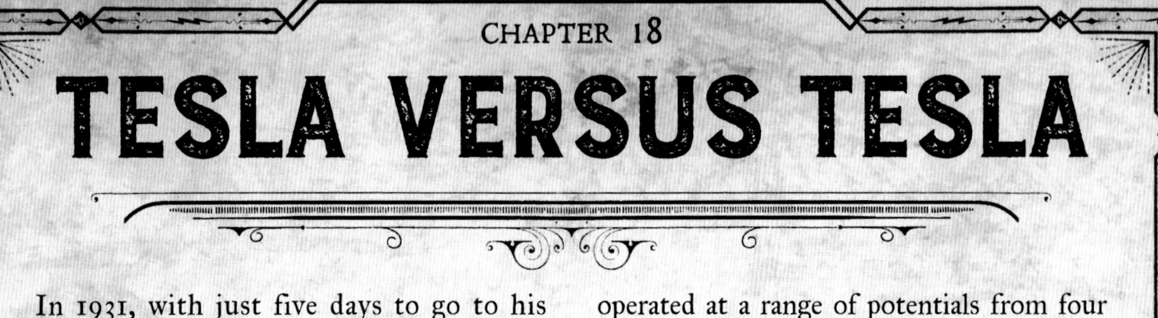

CHAPTER 18

TESLA VERSUS TESLA

In 1931, with just five days to go to his seventy-fifth birthday, Tesla said that he would soon announce "by far the most important discovery" of his long career. It was one of many that began to dent his credibility.

"It will throw light on many puzzling phenomena of the cosmos," he said, "and may prove also of great industrial value, particularly in creating a new and virtually unlimited market for steel."[1]

Again, he was tantalizingly vague when it came to the details:

> *I can only say at this time that it will come from an entirely new and unsuspected source, and will be for all practical purposes constant, day and night, and at all times of the year. The apparatus for capturing the energy and transforming it will partake both mechanical and electrical features, and will be of ideal simplicity. At first the cost may be found too high, but this obstacle will be overcome. Moreover, the installment will be, so to speak, indestructible, and will continue to function for any length of time without additional expenditures.*[2]

THE ILLUSION OF ATOMIC ENERGY

Having rejected Einstein's theory of relativity, Tesla argued that nuclear energy was consequently a pipedream.

"The idea of atomic energy is illusionary," he said, "but it has taken so powerful a hold on the minds that, although I have preached against it for twenty-five years, there still are some who believe it to be realizable."[3]

However, Tesla claimed to have disintegrated atoms using the high-potential vacuum tube he had developed in 1896. This, he said, was one of his best inventions. It operated at a range of potentials from four million to eighteen million volts. Since then he had developed an apparatus that would work at fifty million volts, producing results of great scientific importance.

"But as to atomic energy, my experimental observations have shown that the process of disintegration is not accompanied by a liberation of such energy as might be expected from present theories,"[4] he said.

THE ATTACK OF THE DRONES

Tesla told of rocket-ships he had also been designing that, he said, could attain speeds of nearly a mile a second – 3,600 miles an hour (5,793 kilometers per hour) – through the rarefied medium above the stratosphere:

> *I anticipate that such machines will be of tremendous importance in international conflicts in the future. I foresee that in times not too distant wars between various countries will be carried on without a single combatant passing the border. At this very time it is possible to construct such infernal machines which will carry any desired quantity of poisoned gases and explosives, launch them against a target thousands of miles away and destroy a whole city. If wars are not done away with, we are bound to come eventually to this kind of warfare, because it is the most economical means of inflicting injury and striking terror in the hearts of enemies that ever has been imagined. Densely populated countries, like England and Japan, will be at a great disadvantage as compared with those embracing vast territories, such as the United States and Russia.*[5]

Although some of Tesla's ideas in later life can be dismissed as the ravings of a mad scientist, in this case, again, he showed remarkable prescience.

INTERPLANETARY BROTHERHOOD

When *Time* magazine put the aging, eccentric inventor on their cover of the July 20, 1931 issue to celebrate his seventy-fifth birthday, Tesla did not disappoint. He told them of his new invention, the Tesla-Scope, that he could use to send signals to the stars, saying:

> *I think that nothing can be more important than interplanetary communication. It will certainly come some day, and the certitude that there are other human beings in the universe, working, suffering, struggling like ourselves, will produce a magic effect on mankind, and will form the foundation of a universal brotherhood that will last as long as humanity itself.*[6]

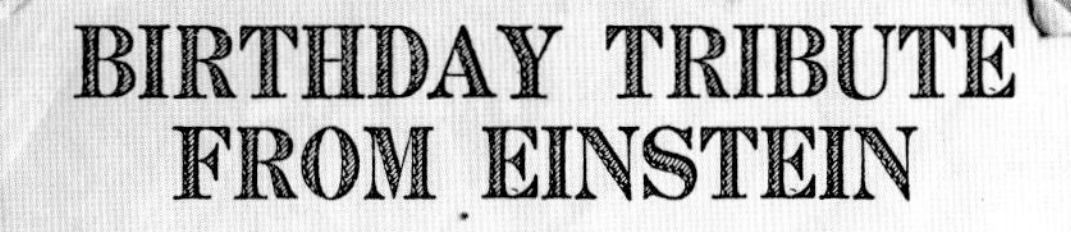

"I'm happy to hear that you are celebrating your seventy-fifth birthday, and that, as a successful pioneer in the field of high-frequency currents, you have been able to witness the wonderful development of this field of technology. I congratulate you on the magnificent success of your life's work."

Albert Einstein[7]

Accolades flooded in. Over a hundred letters of congratulation were received from other scientists including Lee De Forest and Albert Einstein. Notably absent were birthday greetings from Marconi.

PHOTOGRAPHING THOUGHT

At seventy-seven Tesla told a journalist from the *Kansas City Journal-Post* that he expected soon to be able to photograph thoughts:

> *In 1893, while engaged in certain investigations, I became convinced that a definite image formed in thought must, by reflex action, produce a corresponding image on the retina, which might possibly be read by suitable apparatus. It is a mere question of illuminating the same property and taking photographs, and then using the ordinary methods which are available to project the image on a screen. If this can be done successfully, then the objects imagined by a person would be clearly reflected on the screen as they are formed, and in this way every thought of the individual could be read. Our minds would then, indeed, be like open books.*[8]

The cover of *Time* magazine with Nikola Tesla, July 1931.

THE EVIL DEATH RAY

Tesla had inherited a deep hatred of war from his father. Throughout his life, he sought ways to end warfare. Short of that, he thought wars should be fought out between machines. His idea for a death ray began back in the 1890s, when he produced a type of lamp which, with a beam of electrons, could vaporize zirconia or diamonds. And in 1915, he talked of beaming energy from Wardenclyffe, that would "paralyze or kill."

As the storm clouds of war were gathering over Europe again, on July 11, 1934, *The New York Times* carried a headline on its front page which read: *"Tesla, at 78, Bares New Death Beam."* [9]

Tesla said his new invention "will send concentrated beams of particles through the free air, of such tremendous energy that they will bring down a fleet of ten thousand enemy airplanes at a distance of 250 miles (400 km) from a defending nation's borders and will cause armies of millions to drop dead in their tracks." [10]

Tesla said that his death beam would make war impossible by offering every country an "invisible Chinese wall, only a million times more impenetrable." [11] This would make every nation safe from invasion. What's more, they could not be used as offensive weapons as, Tesla said, the death beam "could be generated only from large, stationary and immovable power plants, stationed in the manner of old-time forts at various strategic distances from each country's border ... they could not be moved for purposes of attack." [12]

Submarines would also become obsolete, he said, as methods of detecting them had been perfected to the point where there was no point in submerging. Once a submarine had been located, the death beam could

Tesla's high-frequency invisible electric-ray detection system for locating submerged submarines.

be employed as it would work underwater, though not as well as in air.

Elsewhere he proclaimed that the battleship was doomed. "What happened to the armored knight will also happen to the armored vessel," he said. The money spent on battleships "should be directed in channels that will improve the welfare of the country."[13]

TESLA'S CHARGED-PARTICLE BEAM

The production of a practical death beam involved four new inventions of Tesla's. However, he would not provide details until they had been submitted to the proper scientific authorities. But he reckoned that it would cost no more than $2 million and take only three months to build.

Tesla asked Jack Morgan for the money to build a prototype. Then he tried to deal directly with the British. When they showed no interest, he circulated an elaborate technical paper, called *The New Art of Projecting Concentrated Non-Dispersive Energy Through Natural Media.* This provided the first technical description of a charged-particle beam weapon. Back then this was dismissed as fantasy. However, such things have since been developed. Tesla had even solved one of the key problems of a death ray – how to operate a vacuum chamber needed to accelerate particles with one end open to the atmosphere. He created a "dynamic seal" by directing a high-velocity stream of air at the tip of his gun using a large Tesla turbine.[14]

Interest came from the Soviet Union and, in 1937, Tesla presented a plan to the Amtorg Trading Corporation, in New York City, which handled trade with the Soviet Union. Two years later, in 1939, part of the prototype was tested in the USSR and Tesla received a check for $25,000.[15] But by then, the Soviet Union had allied itself with Nazi Germany.

While Tesla's death beam did not see the light of day in World War II, during the Cold War both the US and the Soviet Union worked on charged-particle beams.

TESLA'S GREATEST HITS

Thirty journalists joined Tesla in 1935, to celebrate his seventy-ninth birthday in the private dining room of the Hotel New Yorker, where he was then staying. He had been thrown out of the Hotel Pennsylvania in 1930, owing $2,000, when other patrons complained of the pigeon droppings. The Hotel New Yorker then supplied a birthday cake with one candle for their honored guest.

Asked what was his greatest feat in the field of engineering, he said: "An apparatus by which mechanical energy can be transmitted to any part of the terrestrial globe."[16]

Tesla called this discovery "tele-geo-dynamics" and admitted that it would "appear almost preposterous." But it would give the world a new means of unfailing communication, provide a new and by far the safest means for guiding ships at sea and into port, furnish a "divining rod" for locating any type of ore beneath the surface of the Earth, give scientists a means of "laying bare the physical conditions of the Earth and enable them to determine all the Earth's physical constants."[17] He had combined cellphones, geo-prospecting and satellite navigation into one amazing apparatus.

His second greatest invention, he said, "will be considered absolutely impossible by any competent electrical engineer." It was a new method of producing direct current without a commutator – something, he said, "that has been considered impossible since the days of Faraday."[18]

"Incredible as it seems," he said, "I have found a solution for this old problem."

Next he came to cosmic rays which, he said, were produced by the force of electrostatic repulsion and consisted of powerfully charged positive particles that come to Earth from the Sun and other stars.

OPENING THE TESLA INSTITUTE

Despite his eccentricities, Tesla's contribution was still recognized. The Tesla Institute was opened in Belgrade, then the capital of Yugoslavia in 1936. A week of celebrations marking the great inventor's eightieth birthday followed – in Belgrade on May 26, 27, and 28, in Zagreb, capital of Croatia, on May 30 and in his native village of Smiljan on June 2, and again on July 12.[19]

At eighty-one, Tesla was awarded the Grand Cordon of the White Eagle, Yugoslavia's highest honor bestowed by King Peter through his regent, Prince Paul, and Czechoslovakia's Order of the White Lion. These were presented at his traditional birthday luncheon at the Hotel New Yorker while Tesla was making his annual announcements of new discoveries to newspapermen.

The following year, he told the press that he had perfected the principle of a new tube that would make it possible to smash atoms and produce cheap radium. They would not have to wait long until he gave them a demonstration, he said.

While he was certain that there was life on other planets, the problem with his equipment, he said, was hitting other moving planets with a needle-point of tremendous energy. But he thought that astronomers could help solve this problem. First they should aim his Tesla Ray at the Moon.

Tesla in the Hotel New Yorker in the late 1930s. After an accident, he was walking with a cane.

RUNNING OUT OF COLLATERAL

While Tesla's mind was as fertile as ever, his financial situation continued to decline. Unable to pay his hotel bill, again, Tesla handed over the "working model" of his death beam as collateral. It was worth $10,000, he said. He also told Jack Morgan that the Russians were keen to buy his death beam to defend themselves against the Japanese. However, he already owed Morgan a great deal of money over his bladeless turbines and no money was forthcoming.

When Gernsback showed him Westinghouse's latest radio set, Tesla saw immediately that the company were flagrantly infringing his wireless patents. He voiced his protest, but was in no position to fight a large corporation.

Eventually, Westinghouse acknowledged Tesla's contribution to the company and paid him $125 a month as a consulting engineer. They also came to an agreement with the Hotel New Yorker where Tesla lived rent free for the rest of his life. In his last years, the Yugoslav government also gave him an honorarium of $7,200 a year.

DYING ALONE

During the latter part of 1942, Tesla became practically a recluse. Physically weak, he retired to bed and permitted no visitors. Hotel staff were not to visit his room unless he summoned them and he refused to listen to any suggestion that they call a doctor. On January 5, 1943, he called a maid to his room and told her that he was not to be disturbed. Nothing was heard from him for three days. Finally, the maid decided to risk his wrath and check up on him. She entered the room in trepidation and found him dead.

The Medical Examiner determined that Tesla had died in his sleep at 10.30 p.m. on Thursday, January 7, 1943, just a few hours before the maid's early morning visit. There were "No suspicious circumstances."[20] The FBI came to take any papers that might aid the war effort.

TRIBUTES AND EULOGIES

The Yugoslav ambassador, Dr. Constantin Fotitch, laid on a state funeral for Tesla at the Cathedral of St. John the Divine. Over two thousand mourners were present, including other scientists and inventors who paid tribute to his intellect and technological achievements. Telegrams of condolence came from Nobel Prize winners, prominent scientists, literary figures, and US government officials. A message from Mrs. Eleanor Roosevelt read: "The President and I are deeply sorry to hear of the death of Mr. Nikola Tesla. We are grateful for his contribution to science and industry and to this country."[21]

Vice President Henry Wallace paid a more personal tribute: "Nikola Tesla, Yugoslav born, so lived his life as to make it an outstanding example of that power

Tesla's funeral at St. John the Divine Cathedral in New York, January 12, 1943. The casket was covered with flags of both the United States and Yugoslavia.

that makes the United States not merely an English-speaking nation but a nation with universal appeal. In Nikola Tesla's death, the common man loses one of his best friends."[22]

Over the radio, New York Mayor Fiorello La Guardia (1882 – 1947) read a eulogy and, after the service, Tesla's body was taken to Ferncliff Cemetery in Ardsley, New York, and was later cremated.[23]

EXTRAORDINARY MAN OF GENIUS

Tesla had always been loved by the popular press for his shocking experiments and outrageous pronouncements. On his death, the *New York Sun* called him "an extraordinary man of genius."

Hugo Gernsback wrote of Tesla's peculiar greatness:

> *We cannot know, but it may be that a long time from now, when patterns have changed, the critics will take a view of history. They will bracket Tesla with Da Vinci, or with our own Mr. Franklin ... One thing is sure. The world, as we run it today, did not appreciate his peculiar greatness.*[24]

The president of RCA David Sarnoff said: "Nikola Tesla's achievements in electrical science are monuments that symbolize America as a land of freedom and opportunity ... His novel ideas of getting the ether in vibration put him on the frontier of wireless. Tesla's mind was a human dynamo that whirled to benefit mankind."

Radio pioneer Edwin Armstrong (1890 – 1954), who went on to sue RCA for infringing his patents, said:

> *Who today can read a copy of The Inventions, Researches and Writings of Nikola Tesla, published before the turn-of-the-century, without being fascinated by the beauty of the experiments described and struck with admiration for Tesla's extraordinary insight into the nature of the phenomena with which he was dealing? Who now can realize the difficulties he must have had to overcome in those early days? But one can imagine the inspirational effect of the book forty years ago on a boy about to decide to study the electrical art. Its effect was both profound and decisive.*[25]

Nine months after Tesla's death the USS *Nikola Tesla* – a Liberty ship, vital to the Allied war effort – was launched in Baltimore.

THE STOLEN SECRETS

The papers from the safe in Tesla's room were lodged with the Office of Alien Property. This was unusual as Tesla was a US citizen. Among them, it was said, were complete plans for his Death Beam and the still unpublished *The New Art of Projecting Concentrated Non-Dispersive Energy Through Natural Media*. This was classified top secret and distributed to Naval Intelligence, the National Defense Research Council, the FBI, MIT, Wright-Patterson Air Force Base – where the V-1 flying bomb was reverse engineered. It is thought that experiments concerning Tesla's Death Beam were conducted there.[26]

TESLA'S AFTERLIFE

In 1943 – a few months after Tesla's death – the US Supreme Court upheld Tesla's radio patent number 645,576, recognizing him as the inventor of radio whose patents had been infringed by Marconi all those years ago.

The "Tesla" became the unit of magnetic flux density. An asteroid and a crater on the Moon has been named after him. The IEEE has presented the Nikola Tesla Award for outstanding contributions to the generation or utilization of electric power annually since 1976. Belgrade International Airport was renamed Belgrade Nikola Tesla Airport in 2006. But, generally in his adopted country of the USA, Tesla has been overlooked.

However, in 2003, Tesla Motors began producing electric vehicles in California, using engines based on Tesla's designs. They are now quoted on the NASDAQ stock exchange.

Tesla Magazine was launched on July 10, 2013. At the same time, the Tesla Science Foundation joined forces with Belgrade's Nikola Tesla Museum to create a traveling exhibition called "Tesla's Wonderful World of Electricity" which toured the US and Canada. Its aim was to get Tesla the recognition he deserves in America. A new movie is being made about his life.

However, few people understand the workings of AC or the mysteries of radio waves and there is some way to go before Tesla gets the public recognition that Edison enjoys. But even if some of his more amazing predictions from later life prove to be correct, the coming years will, undoubtedly, be all about Tesla.

THE FUTURE IS MINE

Being a great man for predicting the future, it is appropriate that the last word on the subject is left with Nikola Tesla. Traveling back to 1904, we can share his vision of "The Transmission of Electrical Energy without Wires" from his March 5 article in *Electrical World and Engineer*. It demonstrates, in his own inimitable style, just how far ahead of his own time he was, and how Tesla's world is really our world of today ... and tomorrow.

But I am hopeful that these great realizations are not far off, and ... When the great truth accidentally revealed and experimentally confirmed is fully recognized, that this planet, with all its appalling immensity, is to electric currents virtually no more than a small metal ball and that by this fact many possibilities, each baffling imagination and of incalculable consequence, are rendered absolutely sure of accomplishment; when the first plant is inaugurated and it is shown that a telegraphic message, almost as secret and non-interferable as a thought, can be transmitted to any terrestrial distance, the sound of the human voice, with all its intonations and inflections, faithfully and instantly reproduced at any other point of the globe, the energy of a waterfall made available for supplying light, heat or motive power, anywhere – on sea, or land, or high in the air – humanity will be like an ant heap stirred up with a stick: See the excitement coming!

Statue of Nikola Tesla standing on an alternator at Niagara Falls. Designed by Les Drysdale, 2006.

Let the future tell the truth, and evaluate each one according to his work and accomplishments. The present is theirs; the future ... is mine.

Nikola Tesla
(1856 - 1943)

Nikola Tesla's Blue Portrait, painted by Hungarian Princess Vilma Lwoff-Parlaghy in 1916.

SOURCES AND CITATIONS

This biography of the Twin Wizards of Electricity – Thomas Edison and Nikola Tesla – has been researched using the archives of one of the most renowned institutions in the world – the British Library in London. Primary source knowledge from publications such as the biography of Edison by Frank Lewis Dyer and Thomas Commerford Martin, published in 1910 and authorized by Edison himself, documents from the Tesla Museum in Belgrade and Tesla's own autobiography, *My Inventions,* published in 1919, have been blended with articles from contemporary electrical scientific journals and newspapers and material from an extensive variety of more recently published books.

The resulting recreation of the life and times of these two great men, is an informative and entertaining introductory text. There are other more academic publications available should the reader wish to delve more deeply.

Books that were especially useful in the preparation of this book are listed below along with the abbreviations shown in square brackets to refer to them in the full list of citations that follows.

SOURCES

Alvarado, Rudolph Valier, *Thomas Edison* (Indianapolis: Alpha Books, 2002) [Alvarado, *Thomas Edison*]

Burgan, Michael, *Nikola Tesla: Physicist, Inventor, Electrical Engineer,* (Minneapolis, MN: Compass Point Books, 2009) [Burgan, *NT: PIEE*]

Carlson, W. Bernard, *Tesla: Inventor of the Electrical Age* (Princeton, NJ: Princeton University Press, 2013) [Carlson, *Tesla*]

Cawthorne, Nigel, *Tesla: The Life and Times of an Electric Messiah* (New York: Chartwell Books, 2014) [Cawthorne, *Electric Messiah*]

Cheney, Margaret, *Tesla: Man Out of Time* (Englewood Cliffs, NJ: Prentice-Hall, 1986) [Cheney, *Man Out of Time*]

Cheney, Margaret & Uth, Robert, *Tesla, Master of Lightning* (New York: Barnes & Noble, 1999) [Cheney & Uth, *Master of Lightning*]

Dyer, Frank Lewis & Martin, Thomas Commerford, *Edison: His Life and Inventions* (New York: Harper & Brothers, 1910) [Dyer & Martin, *Edison*]

Essig, Mark, *Edison and the Electric Chair* (New York: Walker Publishing Company, Inc., 2003) [Essig, *Electric Chair*]

Evenson, A. Edward, *The Telephone Patent Conspiracy of 1876: The Elisha Gray - Alexander Bell Controversy* (North Carolina: McFarland, 2000) [Evenson, *Telephone Patent Conspiracy*]

Israel, Paul, *Edison: A Life of Invention* (New York: Wiley, 1998) [Israel, *Edison: A Life*]

Jonnes, Jill, *Empires of Light: Edison, Tesla, Westinghouse, and the Race to Electrify the World* (New York: Random House, 2003) [Jonnes, *Empires of Light*]

Josephson, Matthew, *Edison* (New York: McGraw-Hill Book Co., 1959) [Josephson, *Edison*]

Klein, Maury, *The Power Makers: Steam, Electricity, and the Men Who Invented Modern America* (New York: Bloomsbury, 2008) [Klein, *Power Makers*]

Martin, Thomas Commerford, *Inventions, Researches and Writings of Nikola Tesla.* (First published 1894, reprinted New York: Barnes &Noble, 1995) [TCM, *IRWNT*]

Moran, Richard, *Executioner's Current: Thomas Edison, George Westinghouse, and the Invention of the Electric Chair* (New York: Alfred A, Knopf, 2002) [Moran, *Executioner's Current*]

Mrkich, Dan, *Nikola Tesla: The European Years.* (Ottawa: Commoners Publishing, 2004) [Mrkich, *Tesla: The European Years*]

O'Neill, John J., *Prodigal Genius: The Life of Nikola Tesla* (First published 1943, reprinted London: Granada Publishing, 1968) [O'Neill, *Prodigal Genius*]

Passer, Harold, *The Electrical Manufacturers, 1875-1900.* (Cambridge, MA: Harvard University Press, 1953) [Passer, *The Electrical Manufacturers*]

Prout, Henry G., *A Life of George Westinghouse* (New York: American Society of Mechanical Engineers, 1921) [Prout, *George Westinghouse*]

Ratzlaff, John T., *Tesla Said* (Millbrae, Tesla Book Co, 1984) [*Tesla Said*]

Seifer, Marc J. *Wizard: The Life and Times of Nikola Tesla, Biography of a Genius* (Secaucus, NJ: Birch Lane Press, 1996) [Seifer, *Wizard*]

Sonneborn, Liz, *The Electric Light: Thomas Edison's Illuminating Invention* (New York: Chelsea House Publishers, 2007) [Sonneborn, *Electric Light*]

Stewart, Daniel Blair, *Tesla: The Modern Sorcerer* (Berkeley, CA: Frog Ltd., 1999) [Stewart, *Modern Sorcerer*]

Tate, Alfred O., *Edison's Open Door: The Life Story of Thomas A. Edison, A Great Individualist* (New York: E.P. Dutton, 1938) [Tate, *Edison's Open Door*]

Tesla, Nikola & Childress, David H., *The Fantastic Inventions of Nikola Tesla* (Stelle, IL: Adventures Unlimited, 1993) [NT & Childress, *Fantastic Inventions of NT*]

Tesla, Nikola & Childress, David H., *The Tesla* Papers (Stelle, IL: Adventures Unlimited, 2000) [NT & Childress, *The Tesla Papers*]

Tesla, Nikola, *My Inventions and Other Writings* (1919) www.teslasautobiography.com [NT, *My Inventions*]

Tesla, Nikola *Lectures, Patents, Articles (1856-1943)*. (Belgrade: Tesla Museum, 1956) [NT, *Lectures, Patents, Articles*]

Valone, Thomas, *Harnessing the Wheelwork of Nature: Tesla's Science of Energy*, 21-23 (Kempton, IL: Adventures Unlimited Press, 2002) [Valone, *Wheelwork of Nature*]

CITATIONS

INTRODUCTION

1. *Tesla Said*, 208
2. Dyer & Martin, *Edison*, 90
3. grammy.com/photos/special-merit-awards-family-of-thomas-edison
4. Cawthorne, *Electric Messiah*, 181

CHAPTER 1

1. NT, *My Inventions*
2. Cawthorne, *Electric Messiah*, 21
3. Carlson, *Tesla*, 68 (citing Tesla in *Complaint's Record on Final Hearing*, Volume 1: *Westinghouse vs. Mutual Life Insurance Company and H.C. Mandeville* [1903]; NT, "Electric Magnetic Motor", US Patent 424,036 [filed May 20, 1889, granted March 25, 1890]; TCM, *IRWNT*, 69)
4. Hunt, Samantha, *The Invention of Everything Else*, 52 (Boston, MA: Houghton Mifflin, 2008)
5. Stewart, *Modern Sorcerer*, 163
6. NT, *My Inventions*
7. *Buffalo New York News*, August 30, 1896
8. *New York Times*, October 19, 1931
9 – 10. NT, *My Inventions*
11. *Century*, February 1894
12. Tate, *Edison's Open Door*, 149
13 – 15. Dyer & Martin, *Edison*, 6-7
16 – 24. Dyer & Martin, *Edison*, 10-13
25. Dyer & Martin, *Edison*, 18
26. Dyer & Martin, *Edison*, 30
27 – 29. Dyer & Martin, *Edison*, 34-35
30. Dyer & Martin, *Edison*, 23
31. Dyer & Martin, *Edison*, 39
32. Dyer & Martin, *Edison*, 42
33. Dyer & Martin, *Edison*, 46-47
34 – 37. Dyer & Martin, *Edison*, 56-59
38. Israel, *Edison: A Life*, 119
39. Cawthorne, *Electric Messiah*, 24
40. Israel, *Edison: A Life*, 142-48
41. Dyer & Martin, *Edison*, 68-71
42. Alvarado, *Thomas Edison*, 249-253
43. Dyer & Martin, *Edison*, 57
44. Josephson, *Edison*, 145-147
45. Josephson, *Edison*, 144
46. Israel, *Edison: A Life*, 425
47. Evenson, *Telephone Patent Conspiracy*, 182-85
48. Dyer & Martin, *Edison*, 56
49 – 53. Dyer & Martin, *Edison*, 66-69
54. Edison, Thomas Alva, *Encyclopædia Britannica* (Chicago: Encyclopaedia Britannica, 2010)
55. Dyer & Martin, *Edison*, 76
56. Higgs, Paget, *The Electric Light In Its Practical Application*, 214. (London, E. & F.N. Spon, 1879)
57. public-domain-content.com/books/Edison/C11P2
58. Israel, *Edison: A Life*, 173-74
59. Magoc, Chris J., *Chronology of Americans and the Environment*, 46 (Santa Barbara, CA: ABC-CLIO, 2011)
60. Dyer & Martin, *Edison*, 119
61. Dyer & Martin, *Edison*, 107
62. Josephson, *Edison*, 255
63. Jonnes, *Empires of Light*, 83-85
64. alevo.com/edisons-1882-pearl-street-station
65. Cawthorne, *Electric Messiah*, 28-30
66. Hood, Bill, *Reflections on Peace and Tranquillity*, 294 (Google Books, 2015)
67. *Scientific American*, January 18, 1879
68. *Dundee Advertiser*, April 11, 1834

CHAPTER 2

1. Carlson, *Tesla*, 17 & 397
2. Burgan, *NT: PIEE*, 20
3. NT, *My Inventions*
4. *Tesla Said*, 283-84
5 – 9. NT, *My Inventions*
10. Seifer, *Wizard*, 13 (citing NT, April 23, 1893)
11. Cheney & Uth, *Master of Lightning*, 9
12 – 17. NT, *My Inventions*
18. Mrkich, *Tesla: The European Years*, 100
19 – 20. NT, *My Inventions*
21. Cheney, *Man Out of Time*, 25
22 – 26. NT, *My Inventions*
27. *New York Times*, May 12, 1938
28. *New York Sun*, July 12, 1937
29 – 31. Jonnes, *Empires of Light*, 105-106
32. *Scientific American*, June 5, 1915
33 – 34. NT, *My Inventions*
35. O'Neill, *Prodigal Genius*, 62
36. Jonnes, *Empires of Light*, 107
37. Cheney & Uth, *Master of Lightning*, 20
38. Tate, *Edison's Open Door*, 149
39. Cheney, *Man Out of Time*, 34

CHAPTER 3

1. NT, *My Inventions*
2. Seifer, *Wizard*, 40-41 (citing "Tesla Electric Co., [advertisement], *Electrical Review*, September 14, 1886)
3. *Electrical Review*, September 14, 1886
4 – 5. *Tesla Said*, 280
6. NT, *My Inventions*
7. O'Neill, *Prodigal Genius*, 65
8 – 9. Dyer & Martin, *Edison*, 115
10. Jonnes, *Empires of Light*, 113
11. O'Neill, *Prodigal Genius*, 65
12. Seifer, *Wizard*, 42 (citing *Electrician and Electrical Engineer*, 1886)
13 – 15. *Electrical Experimenter*, March 19, 1919
16 – 17. Jonnes, *Empires of Light*, 115
18. Carlson, *Tesla*, (citing Testimony in Complainant's Record on Final Hearing, Volume 1: *Testimony, Westinghouse vs Mutual Life Insurance Company and H.C. Mandeville*, 1903)
19. Carlson, *Tesla*, 84-85 (citing NT 78. 21 "Defendant's Brief, Derivation Electric Motor", in *Westinghouse Electric and Manufacturing Company vs. Dayton Fan and Motor Company*, 1900)

20. NT, *My Inventions*
21. O'Neill, *Prodigal Genius*, 67
22. TCM, *IRWNT*, 106
23. Jonnes, *Empires of Light*, 156

CHAPTER 4

1. Prout, *George Westinghouse*, 5
2. Jonnes, *Empires of Light*, 119-20
3. Leupp, Francis E., *George Westinghouse: His Life and Achievements*, 287 (Boston, 1919)
4. Prout, *George Westinghouse*, 293
5. Jonnes, *Empires of Light*, 123
6. Carlson, *Tesla*, 90; also Klein, *Power Makers*; Davis, L.J., Fleet Fire: Thomas Edison and the Pioneers of the Electric Revolution, *Material History Review*, Volumes 33-37, 25
7. Passer, *The Electrical Manufacturers*, 132

8 – 9. Ferris, George T., ed., *Our Native Land*, 93 (New York: Appleton, 1886)

10 – 12. Jonnes, *Empires of Light*, 130-132 (citing Reginald Belfield, "Westinghouse and Alternating Current", George Westinghouse: Anecdotes and Reminiscences.)

CHAPTER 5

1. *Buffalo Commercial Advertiser*, November 27, 1886
2. Josephson, *Edison*, 346 (citing Edison. Memorandum to E.H. Johnson, 1886, Siemens and Halske's Report on "Z.B.D." AC system.)
3. *New York Daily Tribune*, April 17, 1888
4. *New York Times*, May 12, 1888
5. Sonneborn, *Electric Light*, 75
6. Brandon, Craig, *The Electric Chair: An Unnatural American History*, 58 (North Carolina: McFarland, 1999)
7. Axelrod, Alan, *Profiles in Folly*, 196 (New York: Sterling, 2008)
8. *A Warning from the Edison Electric Light Co.* (New York, 1887)
9. Jonnes, *Empires of Light*, 131 (citing Drew, Bernard A. & Chapman, Gerard, "William Stanley Lighted a Town and Powered an Industry," *Berkshire Today* 6, no 1 [fall 1985], 8)

CHAPTER 6

1. NT, *My Inventions*
2. NT & Childress, *Fantastic Inventions of NT*, 17-38
3. *Science*, May 16, 1958
4. TCM, *IRWNT*, 9

5 – 6. *The Essential Tesla*, 10-11 (Radford, VA: Wilder Publications, 2007)

7. NT, *Lectures, Patents, Articles*, L-12
8. Passer, *The Electrical Manufacturers*, 136
9. *Electrical Engineer*, Volume 7, 277
10. Jonnes, *Empires of Light*, 159
11. Passer, *The Electrical Manufacturers*, 277
12. Jonnes, *Empires of Light*, 160
13. Passer, *The Electrical Manufacturers*, 277
14. Jonnes, *Empires of Light*, 160-161
15. Seifer, *Wizard*, 49-50
16. Passer, *The Electrical Manufacturers*, 278
17. O'Neill, *Prodigal Genius*, 83
18. O'Neill, *Prodigal Genius*, 49

19 – 20. *Electrical World*, March 21, 1914

CHAPTER 7

1. *New York Post*, June 5, 1888
2. Sonneborn, *Electric Light*, Ch 7
3. Jonnes, *Empires of Light*, 166-67
4. Moran, *Executioner's Current*, 56
5. Wiggins, Arthur W. & Harris, Sidney *The Joy of Physics*, 257 (Amhurst, NY: Prometheus Books, 2007)
6. Moran, *Executioner's Current*, 56

7 – 10. *Electrical Engineer*, August 1888

11. McPherson, Stephanie Sammartino, *War of the Currents*, 37 (Breckenridge, CO: 21st Century Books, 2012)
12. Essig, *Electric Chair*, 145

13 – 17. *New York Times*, July 31, 1888

18. Moran, *Executioner's Current*, 99
19. *New York Times*, July 31, 1888
20. Moran, *Executioner's Current*, 99

21 – 23. *New York Morning Sun*, August 4, 1888

24. Klein, *Power Makers*, 265
25. *Electrical Engineer*, September 1888
26. *New York Evening Post*, December 10, 1888

CHAPTER 8

1. Moran, *Executioner's Current*, 103
2. Jonnes, *Empires of Light*, 176
3. Essig, *Electric Chair*, 153
4. *Electrical World*, December 15, 1888
5. Jonnes, *Empires of Light*, 176
6. Essig, *Electric Chair*, 153
7. *New York Times*, December 6, 1888
8. Essig, *Electric Chair*, 155 (citing Medico-Legal Journal)
9. *New York Daily Tribune*, December 10, 1888

10 – 11. *New York Evening Post*, December 12, 1888

12 – 13. *New York Daily Tribune*, December 18, 1888

14. *Electrical Review*, June 30, 1888
15. *Buffalo Evening News*, March 30, 1889
16. Moran, *Executioner's Current*, 160 (citing The People of New York State, Ex. Rel. William Kemmler vs Charles F. Durston, Court of Appeals, Buffalo, New York, 1890)
17. Moran, *Executioner's Current*, 162 (citing Kemmler vs Durston)

18 – 19. Moran, *Executioner's Current*, 163 (citing Kemmler vs Durston)

20. *New York Evening Post*, May 14, 1889
21. Moran, *Executioner's Current*, 169 (citing Kemmler vs Durston)
22. Moran, *Executioner's Current*, 179 (citing Kemmler vs Durston)
23. Moran, *Executioner's Current*, 180 (citing Kemmler vs Durston)
24. *Electrical Review*, October 16, 1889
25. *New York Tribune*, August 7, 1890
26. Moran, *Executioner's Current*, 183 (citing Kemmler vs Durston)
27. *New York Sun*, August 25, 1889

28 – 29. Moran, *Executioner's Current*, 185 (citing Kemmler vs Durston); see also Essig, *Electric Chair*, 210-11

30. Moran, *Executioner's Current*, 188 (citing Kemmler vs Durston, Supreme Court, 1889)
31. Moran, *Executioner's Current*, 197 (citing Commentary, May 1963)
32. *New York Tribune*, May 1, 1890
33. *New York Tribune*, August 9, 1890
34. Essig, *Electric Chair*, 161

35 – 36. *New York Times*, August 7, 1890

37. *Buffalo Evening News*, August 6, 1890
38. *New York Times*, August 7, 1890
39. *Auburn Daily Advertiser*, August 11, 1890
40. *New York Times*, August 7, 1890
41. Moran, *Executioner's Current*, 15

42 – 43. *New York Times*, August 7, 1890

44. *Electrical Review*, August 16, 1890
45. Banner, Stuart, *The Death Penalty*, 186 (Cambridge, MA: Harvard University Press, 2003)

46 – 50. *New York Times*, August 7, 1890

51. *New York World*, November 29, 1929
52. *New York Times*, August 6, 1890
53. Essig, *Electric Chair*, 247
54. *New York Times*, August 7, 1890

CHAPTER 9

1. *Electrical World*, September 20, 1924
2. *Electrical Review*, August 14, 1896
3. O'Neill, *Prodigal Genius*, 77

4 – 5. Dyer & Martin, *Edison*, 244

6. *Electrical Engineering*, May 1934
7. Josephson, *Edison*, 354

8. Strouse, Jean *Morgan: American Financier*, 312 (New York: Random House, 1999)
9. Jonnes, *Empires of Light*, 225
10. Lomas, Robert, *The Man Who Invented the Twentieth Century*, 148 (London: Headline, 1999)
11. O'Neill, *Prodigal Genius*, 290
12 – 13. O'Neill, *Prodigal Genius*, 81-82
14. Lamme, B., *Benjamin Garver Lamme: An Autobiography*, 60 (New York: G.P. Putnam, 1926)
15. NT, *Lectures, Patents, Articles*, L-15
16. TCM, *IRWNT*, 173
17. TCM, *IRWNT*, 189
18. TCM, *IRWNT*, 196
19. TCM, *IRWNT*, 196-97
20. *Electrical World*, May 30, 1891
21. *Harper's Weekly*, July 11, 1891
22. Israel, *Edison: A Life*, 334
23. Jonnes, *Empires of Light*, 233-35
24. Jonnes, *Empires of Light*, 237-38
25. *Electrical Engineer*, July 22, 1891
26. Tate, *Edison's Open Door*, 260-61
27. *New York Times*, February 21, 1892
28. *Electrical Engineer*, February 17, 1892
29. Tate, *Edison's Open Door*, 278-79
30. NT, *My Inventions*
31 – 32. *Electrical Review*, March 19, 1892
33. *Electrical Review*, April 9, 1892
34. Carlson, *Tesla*, 193
35. Seifer, *Wizard*, 99
36. *Electrical Experimenter*, May 1919
37. NT, *My Inventions*
38. *New York World*, April 13, 1930
39. TCM, *IRWNT* 318-20
40. *Electrical Engineer*, June 28, 1893
41. Anderson, Leland (ed) *Nikola Tesla on His Work with Alternating Currents*, XV (Breckenridge, CO: 21st Century Books, 2002)

CHAPTER 10

1. Leupp, Francis E., *George Westinghouse: His Life and Achievements*, 163 (Boston, MA: Little Brown, 1919)
2. *Daily Interocean*, May 17, 1892
3. Klein, *Power Makers*, 312-13
4. Jonnes, *Empires of Light*, 257 (citing McClelland, E.S., "Notes on My Career with Westinghouse," 5-6)
5. Jonnes, *Empires of Light*, 258-61
6. *Electrical Engineering*, August 1943
7. Carlson, *Tesla*, 161 (citing Tesla to Westinghouse, September 12, 1892)
8. Jonnes, *Empires of Light*, 261 (citing Lamme, B., *Benjamin Garver Lamme*, 61)
9. TCM, *IRWNT*, 319-20
10. Cameron, William, *The World's Fair*, 318 (New Haven, CT: James Brennan & Co, 1894)
11. Jonnes, *Empires of Light*, 270
12. Jonnes, *Empires of Light*, 264-65 (citing J.P. Barrett, J.P. in G.R. Davis, G.R., *World's Columbian Exposition*, 301)
13. Barrett, J.P. *Electricity at the Columbian Exposition*, 168-69 (Chicago: R.R. Donnelly, 1894)
14. Baldwin, Neil, *Edison: Inventing the Century*, 211-12 (Chicago: University Press, 1995)
15. *Electrical Experimenter*, March 1919
16. *Chicago Tribune*, June 2, 1893
17. *Electrical Engineer*, Volume 16, 153
18. Barrett, J.P. *Electricity at the Columbian Exposition*, 169 (Chicago: R.R. Donnelly, 1894)
19 – 20. *Chicago Tribune*, August 26, 1893

CHAPTER 11

1. *World*, July 22, 1894
2. *Review of Reviews*, March 1894
3. *Century*, February 1894
4. Seifer, *Wizard*, 130
5. *New York World*, July 22, 1894
6. Cheney & Uth, *Master of Lightning*, 52
7 – 8. *New York World*, July 22, 1894
9. *English Mechanic and World of Science*, March 8, 1907
10. *Electrical Engineer*, June 1892
11. Jonnes, *Empires of Light*, 293 (citing Coleman Sellers, *Report on Dynamos*, 25)
12. Jonnes, *Empires of Light*, 293 (citing *Rowland's Final Report*, March 1, 1893, 48)
13 – 14. Jonnes, *Empires of Light*, 293 (citing Coleman Sellers, *Report on Dynamos*, 25)
15 – 16. Jonnes, *Empires of Light*, 294
17. Jonnes, *Empires of Light*, 294 (citing correspondence Tesla-Adams, March-May 1893)
18. Adams, Edward Dean, *Niagara Power*, Vol. 2, 256
19. Jonnes, *Empires of Light*, 298
20. Passer, *The Electrical Manufacturers*, 289
21. Jonnes, *Empires of Light*, 303 (citing Sellers)
22. Passer, *The Electrical Manufacturers*, 289
23. Prout, *George Westinghouse*, 1152-53

CHAPTER 12

1. *New York Sun*, March 14, 1895
2. *New York Times*, March 14, 1895
3. *Troy Press*, April 20, 1895
4. Seifer, *Wizard*, 149 (citing Ratzlaff J., & Anderson, L., *Tesla Bibliography*, 36)
5. *Philadelphia Press*, June 24,1895
6. *New York Times*, July 9, 1895
7. *New York Times*, August 27, 1895
8. Jonnes, *Empires of Light*, 321-22
9. *Western Electrician*, August 1, 1896
10. *Niagara Falls Gazette*, July 20, 1896
11. *Harper's Weekly*, July 21, 1906
12. *Western Electrician*, August 1, 1896
13. *Daily Cataract* (Niagara Falls), July 20, 1896
14. Cheney & Uth, *Master of Lightning*, 61
15. *Buffalo Morning Express*, January 13, 1897
16. *Buffalo Evening News*, January 13, 1897
17. *Electrical World*, February 6, 1897
18. *Buffalo Evening News*, January 12, 1897

CHAPTER 13

1. Skrabec, Quentin R., *George Westinghouse: Gentle Genius*, 195 (New York: Algora Publishing, 2007)
2. Prout, *George Westinghouse*, 187
3. Prout, *George Westinghouse*, 204
4. George Westinghouse, 1846 – 1914 (Westinghouse Electric Corporation), 32
5. "A Challenge to the World" (Westinghouse Educational Foundation), 191
6. Prout, *George Westinghouse*, 188
7. Prout, *George Westinghouse*, 206
8 – 11. Prout, *George Westinghouse*, 197-99
12. *New York Times*, October 24, 1907
13. Skrabec, Quentin R., *George Westinghouse: Gentle Genius*, 207 (New York: Algora Publishing, 2007)
14. Prout, *George Westinghouse*, 200

CHAPTER 14

1. Josephson, *Edison*, 367-79
2. Gracyk, Tim & Hoffman, Frank, Tim *Popular American Recording Pioneers: 1895-1925*, 67 (Binghamton, NY: Haworth Press, 2000)
3. *Scientific American*, June 5, 1880
4. Dyer & Martin, *Edison*, 486
5 –10. Alvarado, *Thomas Edison*, 184-88

11. Brayer, Elizabeth, *George Eastman: A Biography*, 109 (Rochester, NY: University of Rochester Press, 2006)
12. Alvarado, *Thomas Edison*, 189
13 – 14. *Harper's Weekly*, Volume 35, 446

CHAPTER 15

1. Alvarado, *Thomas Edison*, 191
2. Burgan, Michael, *Thomas Alva Edison: Great American Inventor*, 76 (Minneapolis, MN: Compass Point Books, 2006)
3. Pizzitola, Louis, *Hearst Over Hollywood*, 54 (New York: Columbia University Press, 2002)
4. Josephson, *Edison*, 396
5. Alvarado, *Thomas Edison*, 194
6. *New York Times*, April 24, 1896
7. Alvarado, *Thomas Edison*, 194-95
8 – 9. Josephson, *Edison*, 400-1
10. Alvarado, *Thomas Edison*, 197-98
11. Josephson, *Edison*, 404-31
12. Alvarado, *Thomas Edison*, 200
13. *New York Times*, June 24, 1923
14. Alvarado, *Thomas Edison*, 207
15. Curcio, Vincent, *Henry Ford*, 30 (New York: Oxford University Press, 2013)
16. Josephson, *Edison*, 479
17. Alvarado, *Thomas Edison*, 209
18. Shachtman, Tom, *The Day America Crashed*, 20 (New York: Putnam, 1979)
19. Josephson, *Edison*, 479
20. Adair, Gene, *Thomas Alva Edison: Inventing the Electric Age*, 126 (New York: Oxford University Press, 1997)
21. Josephson, *Edison*, 480
22. Alvarado, *Thomas Edison*, 211-12
23. Josephson, *Edison*, 481
24. Alvarado, *Thomas Edison*, 218
25. presidency.ucsb.edu/ws/?pid=22856

CHAPTER 16

1. Seifer, *Wizard*, 189
2. Benson, Allan L., "Nikola Tesla Dreamer" *The World Today*, 1763 (Originally published 1912)
3. *Century*, June 1900
4. *Electrical Engineer*, November 24, 1898
5. *Century*, June 1900
6. Carlson, *Tesla*, 257-58
7. *Colorado Springs Evening Telegraph*, June 2, 1899
8. Seifer, *Wizard*, 218 (citing Tesla to Astor, September 10, 1900)
9. *New York Sun*, January 3, 1901
10. Seifer, *Wizard*, 227 (citing Tesla to U.S. Navy, September 27, 1899)
11. Howeth, L.S., *History of Communications-Electronics in the United States Navy*, Ch4 (Washington DC: Bureau of Ships and Office of Naval History, 1963)
12. O'Neill, *Prodigal Genius*, 186-87
13. *Tesla Journal*, 26-29
14. *Electrical World & Engineer*, January 7, 1905
15. Cheney, *Man Out of Time*, 158; O'Neill, *Prodigal Genius*, 198
16. *Electrical Review*, January 12, 1901
17. Seifer, *Wizard*, 267
18. NT, *My Inventions*
19. Cheney, *Man Out of Time*, 203
20 – 21. *New York Times*, January 14, 1902
22. Carlson, *Tesla*, 339 (citing Tesla to Morgan, January 9, 1902)
23. Cheney & Uth, *Master of Lightning*, 100
24. *Electrical World and Engineer*, March 5, 1904
25. *New York Times*, October 16, 1907
26. *Wireless Telegraphy and Telephony*, 1908

CHAPTER 17

1. *New York Times*, March 11, 1908
2. *New York Times*, June 8, 1908
3. *The Bee*, Danville, Virginia, March 2, 1928
4. *New York Herald*, October 15, 1911
5. *New York Herald Tribune*, October 15, 1911
6. Carlson, *Tesla*, 373 (citing Seifer, *Wizard*, 362-66)
7. O'Neill, *Prodigal Genius*, 173
8. Seifer, *Wizard*, 347-49 (citing correspondence)
9. NT & Childress, *The Tesla Papers*, 120
10. *Electrical Review and Western Electrician*, May 20, 1911
11. *New York Times*, May 16, 1911
12. *New York Times*, August 18, 1912
13. Seifer, *Wizard*, 370 (citing correspondence)
14. Jolly, W. P., *Marconi*, 225 (Stein and Day, 1972)
15. Anderson, Leland (ed) *Nikola Tesla on His Work with Alternating Currents*, 105 (Breckenridge, CO: 21st Century Books, 2002)
16. Seifer, *Wizard*, 377 (citing Lloyd Scott, Naval Consulting Board of the US, Washington, DC: Government Printing Office, 1920)
17 – 18. *New York Times*, October 3, 1915
19. *New York Times*, December 8, 1915
20. Cheney, *Man Out of Time*, 213
21. NT, *My Inventions*
22. O'Neill, *Prodigal Genius*, 229-31
23. *Electrical Experimenter*, September 1917
24. Seifer, *Wizard*, 392-94 (citing correspondence)
25. Cheney, *Man Out of Time*, 223 (citing Tesla, *My Inventions*)
26. Seifer, *Wizard*, 428 (citing Tesla to Carl Laemmle, July 15, 1937; Gabler, Neal, *An Empire of Their Own*, 58)
27. *International Science and Technology*, November 1963
28. O'Neill, *Prodigal Genius*, 293
29. Potter, Adrian, *FBI Report of Friends of Soviet Russia*, 1921-1923
30 – 31. *New York World*, November 21, 1926
32. *Philadelphia Public Ledger*, November 2, 1933

CHAPTER 18

1 – 2. *Time*, July 20, 1931
3 – 5. *New York Times*, July 5, 1931
6. *Time*, July 20, 1931
7. Burgan, *NT: PIEE*, 87
8. *Kansas City Journal-Post*, September 10, 1933
9 – 12. *New York Times*, July 11, 1934
13. *Baltimore Sun*, July 12, 1940
14. tfcbooks.com/tesla/1935-00-00.htm
15. *Newsweek*, May 19, 1986
16 – 18. *New York Times*, July 11, 1935
19. Teslasociety.com/ntinn.htm
20. Cheney, *Man Out of Time*, 265
21 – 23. *New York Times*, January 13, 1943
24. Seifer, *Wizard*, 444
25. Gernsback, Hugo, "NT: Father of Wireless, 1857-1943," *Radio Craft*, February 1943
26. Seifer, *Wizard*, 458-59 (citing Conroy, E.E. to J. Edgar Hoover, FBI, October 17, 1945)

INDEX

This edition published in 2016 by
Chartwell Books
an imprint of The Quarto Group
142 West 36th Street, 4th Floor
New York, New York 10018
USA
www.QuartoKnows.com

ISBN-13: 978-0-7858-3378-9

Printed in China

10 9 8 7 6 5 4 3

PICTURE CREDITS

Front cover: Sarony 1890 / *Electrical Review* 1899 / nps.gov/edis / Bachrach 1922.

Back cover: *New York World* 1894 / © George Eastman House/Getty Images / Edison US patent 263,878 1882.

Patent diagrams: www.uspto.gov / www.google.com/patents : 18 Edison US patent 90,646 1869 / 41 Tesla US patent 396,121 1889 / 44 Tesla US patent 428,057 1890 / 159 Tesla US patent 613,809 1898 / 166 Tesla US patent 1,655,114 1928 / 167 Tesla US patent 1,061,206 1913

Internal: 3 Sarony 1890 / © GL Archive/Alamy / 4 Bachrach Studios 1922 / 6 © Pictorial Press Ltd/Alamy / 7 nps.gov/edis / 10 Sarony 1884 / 11 tesla-universe / 12 © GL Archive/Alamy / 13 Smithsonian NMAH / 15 edison.rutgers.edu / 16 nps.gov/edis / 17 edison.rutgers.edu / 19 nps.gov/edis / 20 *Popular Science Monthly* 1878 / 21 Universal History Archive 1877 / 22 Brady-Handy Photograph Collection 1878 / 24 APT/Building Technology Heritage Library / 25 nps.gov/edis / 26 Edison and Swan Electric Light Company/lamptech.co.uk / 27 Joseph Wilson Swan 1910 / 28 National Portrait Gallery, London / 29 flickriver.com / 30 Smithsonian NMAH / 31 tesla-museum.org / 32 wordpress.com / Hippolyte Fontaine, *Electric Lighting* 1878 / 35 edisontechcenter.org / 36 untappedcities.com / 38 New York Historical Society / 39 *Harper's Weekly* 1882 / 40 untappedcities.com / 42 *Electrical Experimenter* Vol 22 / 48 heinzhistorycenter.org/collections/westinghouse / 49 William Wallace Wood, *The Westinghouse E-T Air Brake Instruction Pocket Book* 1909 / 50 Franklin Institute 1885 / 52 © AS400 DB/Corbis / 53 Louis Figuier, *Les Merveilles de la science,* 1891, Tome 6, djvu / 54 © North Wind Picture Archives/Alamy / 55 Emile Desbeaux, *Physique Populaire* 1891 / 57 *History of the Transformer* 1889 / 58 Thomas Phillips 1842 / 60 Science Museum / 61 Elihu Thomson 1880 / 65 Joseph G. Gessford/loc.gov / 69 *St Louis Republic* 1903 / 70 *Electric Railway Review* 1895 / 74 *Scientific American* 1888 / 76 Science Photo Library / 77 archive.org/stream/famousamericans 1901 / 78 *Le Petit Parisien* 1890 / 81 © AF archive/Alamy / 82 Sarony 1880 / 83 Sarony, *Electrical Experimenter* 1919 / 84 BnF, Estampes-et-Photographie 1889 / 85 Marine Corps Archives & Special Collections / 86 Bibliothek allgemeinen und praktischen Wissens für Militäranwärter Band III, 1905 / 87 G. Brogi, Florenz / 90 *Electrical World* 1891 / 91 tesla-universe / 93 National Portrait Gallery, Washington / 94 Eye of Sauron / tesladownunder.com / 97 Robert Krewaldt 1894 / 98 *Pearson's Magazine* 1899 / 99 Paul Downey@flickr.com /100 Life Photo Archive 1901 / 102 *Electrical Review* 1899 / 106 Chicago History Museum / 108 tesla-universe / 110 heinzhistorycenter.org/collections/westinghouse / 112 Newspaper article *c.*1894 / 113 1893worldsfair.tumblr.com / 114 teslasociety.com / 116 *Century Magazine* April 1895 / 117 Arthur Brisbane 1906/loc.gov / 118 Edward Dean Adams, *Niagara Power* 1918, / 119 AIEE, 1904 /120 *The World's Work* 1902 / 121 Fort Lewis College Center for Southwest Studies / 124 Detroit Publishing Company 1906 / 126 Sarony 1890 / 128 & 129 *Popular Science Monthly* Vol 73 Oct 1908 / 135 Alfred John West 1897 / 137 © World History Archive/Alamy / 140 nps.gov/edis / 141 © James Pintar/Alamy / 142 Muybridge_horse_gallop.jpg ("Daisy") *c.*1878 / 142 *La Nature* n°464 du 22 avril 1882 / 143 © AF Fotografie/Alamy / 144 *Dickson Greeting* 1891 / 146 *Scientific American* 1889 / 147 institut-lumiere.org / 148 *New York Herald* 1896 / 150 © Hulton Archive/Getty Images / 151 *Scientific American* 1896 / 153 greenbiz.com / 155 Henry Ford Museum & Greenfield Village / 158 officemuseum.com / 159 & 160 *Century Magazine* 1900 / 162 tesla-universe / 164 hipertextual.com / 169 *Electrical Experimenter* 1915 / 172 tesla-universe / 173 *The Experimenter* magazine, March 1925 / 174 Waltham 1102 Speedometer 1922 Ad / 177 *Time* 1931 / 178 *Electrical Experimenter* Vol 21, p.122 /180 merlinthegrey.deviantart.com / 181 teslasociety.com / 183 © Washington Imaging/Alamy / 185 Nordsee Museum